Pressure Vessels

Design and Practice

Mechanical Engineering Series

Frank Kreith - Series Editor

Published Titles

Distributed Generation: The Power Paradigm for the New Millennium
Anne-Marie Borbely & Jan F. Kreider
Elastoplasticity Theory
Vlado A. Lubarda
Energy Audit of Building Systems: An Engineering Approach
Moncef Krarti
Engineering Experimentation
Euan Somerscales
Entropy Generation Minimization
Adrian Bejan
Finite Element Method Using MATLAB, 2nd Edition
Young W. Kwon & Hyochoong Bang
Fluid Power Circuits and Controls: Fundamentals and Applications
John S. Cundiff
Fundamentals of Environmental Discharge Modeling
Lorin R. Davis
Heat Transfer in Single and Multiphase Systems
Greg F. Naterer
Introductory Finite Element Method
Chandrakant S. Desai & Tribikram Kundu
Intelligent Transportation Systems: New Principles and Architectures
Sumit Ghosh & Tony Lee
Mathematical & Physical Modeling of Materials Processing Operations
Olusegun Johnson Ilegbusi, Manabu Iguchi & Walter E. Wahnsiedler
Mechanics of Composite Materials
Autar K. Kaw
Mechanics of Fatigue
Vladimir V. Bolotin
Mechanics of Solids and Shells: Theories and Approximation
Gerald Wempner & Demosthenes Talaslidis
Mechanism Design: Enumeration of Kinematic Structures According to Function
Lung-Wen Tsai
Nonlinear Analysis of Structures
M. Sathyamoorthy
Practical Inverse Analysis in Engineering
David M. Trujillo & Henry R. Busby
Pressure Vessels: Design and Practice
Somnath Chattopadhyay
Principles of Solid Mechanics
Rowland Richards, Jr.
Thermodynamics for Engineers
Kau-Fui Wong
Viscoelastic Solids
Roderic S. Lakes

Pressure Vessels

Design and Practice

Somnath Chattopadhyay

CRC PRESS

Boca Raton London New York Washington, D.C.

FIRST INDIAN REPRINT, 2010

Library of Congress Cataloging-in-Publication Data

Chattopadhyay, Somnath.
Pressure vessels : design and practice / Somnath Chattopadhyay.
p. cm. – (Mechanical engineering 25)
Includes bibliographical references and index.
ISBN 0-8493-1369-4
1. Pressure vessels–Design and construction. 2. Structural engineering. I. Title. II. Series: Mechanical engineering series (Boca Raton, Fla.); 25.

TA660.T34C49 2004
681′.76041–dc22 2004051972

Direct all inquiries to CRC Press LLC, 2000 N.W. Corporate Blvd., Boca Raton, Florida 33431.

Visit the CRC Press Web site at www.crcpress.com

International Standard Book Number 0-8493-1369-4
Library of Congress Card Number 2004051972
Printed and bound in India by Nutech Photolithographers, New Delhi

FOR SALE IN SOUTH ASIA ONLY.

Preface

The purpose of this book is to outline the principles that form the basis of design of pressure vessels for power generation and process equipment. It is intended for senior and graduate engineering students as well as for practicing engineers who are interested in learning the basic principles associated with pressure vessel design as they are outlined in the various codes, such as the American Society of Mechanical Engineers Boiler and Pressure Vessel Code. Although some of the topics in pressure vessels are treated in courses on mechanics of materials and mechanical design, a textbook covering design of pressure vessels for graduate and undergraduate students is not available; there are, however, reference texts that outline some of the design rules and some that provide guidance on how the code equations are to be used.

This book has evolved from my long and diverse experience working with pressure vessels and was motivated by a course entitled "Special Topics in Mechanical Engineering" offered during my tenure at the American University in Cairo. I always felt a strong need for such a text in which the principles associated with pressure vessel design are highlighted so that the reader can trace the origin of the design equations used in the construction codes for pressure vessels and obtain a physical insight into the design process. Otherwise the approach is similar to that of following a cookery book, which can sometimes be quite counterproductive.

Design of pressure vessels is important not only for mechanical and structural engineers, but also for chemical engineers involved in the design of chemical and petrochemical equipment, and for material engineers involved with material selection and fabrication issues. This book covers components such as cylindrical shells, formed heads and transition sections, covers and flanges, openings and nozzles, bolted flanges, and vessel supports. The structural design criteria are first introduced and are then followed by stress categories and stress limits as employed in the various design codes. The final chapter discusses a few case studies that will be helpful to designers. The topics of solid mechanics, fatigue and fracture mechanics, and stability and limit analysis are presented as separate appendices that can also be used as a background material for the main chapters. Several example problems are included in the book.

Som Chattopadhyay
Bell State University

Acknowledgments

I would like to acknowledge the support and encouragement from a number of people in this effort. I am grateful to Dr. Sam Palusamy of Westinghouse Electric Company, who as the past Chairman of the ASME Code Subgroup on Design and Analysis as well as the Pressure Vessel Research Council Subcommittee on Shells and Ligaments provided some useful insights into the design issues of pressure vessels. I would also like to thank Professor Maher Younan of the American University in Cairo, for laying the foundation for a senior level course on pressure vessels, which I taught at that institution. In addition I am indebted to a number of individuals from Westinghouse for their continuing support. This includes Alan H. Huey, manager of Strategic Marketing Communications, for providing me with color pictures of nuclear plant cutaways which appear in the book. In particular, I am thankful to Robert J. Sterdis, manager of Steam Generator Design and Analysis, who was instrumental in providing the opportunity to work hands-on in the area of pressure vessel design between my academic schedules. I am deeply indebted to my wife Mandira for her continued encouragement and sacrifice. I would like to thank our children, Somak, Parama and Samik for their cooperation and patience.

Som Chattopadhyay
Muncie, IN

Contents

chapter one

Overview of pressure vessels

Contents

1.1 Introduction

Vessels, tanks, and pipelines that carry, store, or receive fluids are called pressure vessels. A pressure vessel is defined as a container with a pressure differential between inside and outside. The inside pressure is usually higher than the outside, except for some isolated situations. The fluid inside the vessel may undergo a change in state as in the case of steam boilers, or may combine with other reagents as in the case of a chemical reactor. Pressure vessels often have a combination of high pressures together with high temperatures, and in some cases flammable fluids or highly radioactive materials. Because of such hazards it is imperative that the design be such that no leakage can occur. In addition these vessels have to be designed carefully to cope with the operating temperature and pressure. It should be borne in mind that the rupture of a pressure vessel has a potential to cause extensive physical injury and property damage. Plant safety and integrity are of fundamental concern in pressure vessel design and these of course depend on the adequacy of design codes.

When discussing pressure vessels we must also consider tanks. Pressure vessels and tanks are significantly different in both design and construction: tanks, unlike pressure vessels, are limited to atmospheric pressure; and pressure vessels often have internals while most tanks do not (and those that do are limited to heating coils or mixers).

Pressure vessels are used in a number of industries; for example, the power generation industry for fossil and nuclear power, the petrochemical industry for storing and processing crude petroleum oil in tank farms as well as storing gasoline in service stations, and the chemical industry (in chemical reactors) to name but a few. Their use has expanded throughout the world. Pressure vessels and tanks are, in fact, essential to the chemical, petroleum, petrochemical and nuclear industries. It is in this class of equipment that the reactions, separations, and storage of raw materials occur. Generally speaking, pressurized equipment is required for a wide range of industrial plant for storage and manufacturing purposes.

The size and geometric form of pressure vessels vary greatly from the large cylindrical vessels used for high-pressure gas storage to the small size used as hydraulic units for aircraft. Some are buried in the ground or deep in the ocean, but most are positioned on ground or supported in platforms. Pressure vessels are usually spherical or cylindrical, with domed ends. The cylindrical vessels are generally preferred, since they present simpler manufacturing problems and make better use of the available space. Boiler drums, heat exchangers, chemical reactors, and so on, are generally cylindrical. Spherical vessels have the advantage of requiring thinner walls for a given pressure and diameter than the equivalent cylinder. Therefore they are used for large gas or liquid containers, gas-cooled nuclear reactors, containment buildings for nuclear plant, and so on. Containment vessels for liquids at very low pressures are sometimes in the form of lobed spheroids or in the shape of a drop. This has the advantage of providing the best possible stress distribution when the tank is full.

The construction of a typical pressure vessel is shown in Figure 1.1. A spherical pressure vessel is shown in Figure 1.2. This is a special pressure

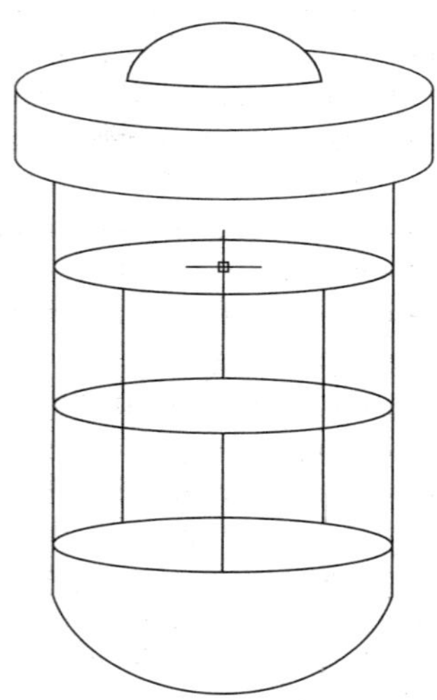

Figure 1.1 Typical pressure vessel.

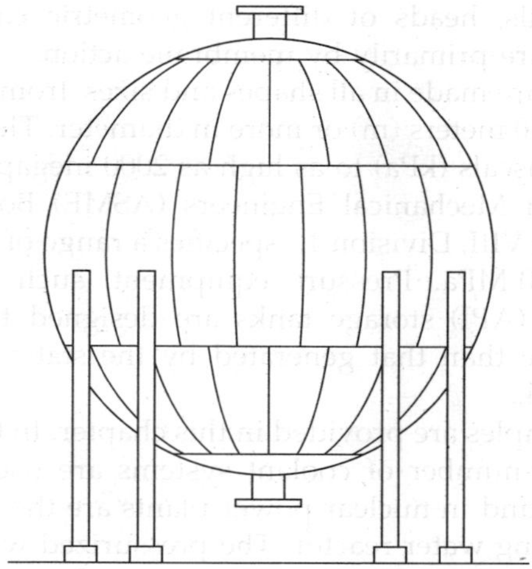

Figure 1.2 Spherical pressure vessel.

vessel and is really a storage sphere. Functionally it acts as a tank because its purpose is to store a fluid. However since it does so at pressures above atmospheric, it can be classified as a pressure vessel. This however does not have internals and operates at atmospheric temperatures. A horizontally supported cylindrical pressure vessel with a hemispherical head and conical transition is shown in Figure 1.3. This consists of a cylindrical main shell, with hemispherical headers and several nozzle connections.

The vessel geometries can be broadly divided into plate- and shell-type configurations. The plate-type construction used in flat covers (closures for pressure vessels and heat exchangers) resists pressure in bending, while the shell-type's membrane action operates in a fashion analogous to what happens in balloons under pressure. Generally speaking the shell-type construction is the preferred form because it requires less thickness (as can be demonstrated analytically) and therefore less material is required for its manufacture. Shell-type pressure components such as pressure vessel and

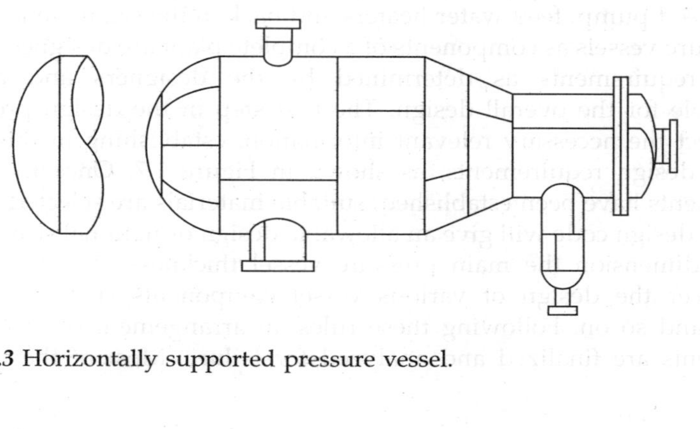

Figure 1.3 Horizontally supported pressure vessel.

heat exchanger shells, heads of different geometric configurations, and nozzles resist pressure primarily by membrane action.

Pressure vessels are made in all shapes and sizes, from a few centimeters (cm) in diameter to 50 meters (m) or more in diameter. The pressure may be as low as 0.25 kilopascals (kPa) to as high as 2000 megapascals (MPa). The American Society of Mechanical Engineers (ASME) Boiler and Pressure Vessel Code, Section VIII, Division 1,[1] specifies a range of internal pressures from 0.1 MPa to 30 MPa. Pressure equipment, such as the American Petroleum Institute (API) storage tanks are designed to restrict internal pressure to no more than that generated by the static head of the fluid contained in the tank.

A few more examples are provided in this chapter. In the area of nuclear power generation a number of coolant systems are used. The two plant cycles most often found in nuclear power plants are the pressurized water reactor and the boiling water reactor. The pressurized water reactor inside the reactor pressure vessel is subjected to a high coolant water pressure. The pressurized water is heated and the pump circulates the water through a heat exchanger (steam generator) where the steam for the turbine is generated. The part of the nuclear power plant containing the reactor coolant is called the primary circuit. Included in the primary circuit is an important vessel called the pressurizer. The coolant volume varies when the load changes require reactor coolant temperature changes, and when this occurs, the pressurizer serves as the expansion tank in the primary system, which allows the water to undergo thermal expansion and contraction keeping the primary circuit pressure nearly constant. If the pressures are allowed to fluctuate too far, steam bubbles might form at the reactor heating surfaces; these bubbles or voids if formed inside the reactor core greatly alter reactor power output. The pressurizer has electric heating elements located low inside to provide the vapor needed to cushion the flowing liquid coolant. All of these items are included in the primary circuit. Figure 1.4 shows a pressurized water reactor (PWR) vessel. A PWR steam generator and a PWR pressurizer are indicated in Figures 1.5 and 1.6, respectively. The rest of the plant is called the secondary circuit. The steam generator produces the steam that passes through the turbine, condenser, condensate pumps, feed pump, feed water heaters and back to the steam generator.

Pressure vessels as components of a complete plant are designed to meet various requirements as determined by the designers and analysts responsible for the overall design. The first step in the design procedure is to select the necessary relevant information, establishing in this way a body of design requirements, as shown in Figure 1.7. Once the design requirements have been established, suitable materials are selected and the specified design code will give an allowable design or nominal stress that is used to dimension the main pressure vessel thickness. Additional code rules cover the design of various vessel components such as nozzles, flanges, and so on. Following these rules an arrangement of the various components are finalized and analyzed for failure. Most of the types of

Figure 1.4 Pressurized water reactor (PWR) pressure vessel.
(Courtesy of Westinghouse Electric Company, Pittsburgh, PA.)

failure relevant to pressure vessel design are stress dependent and therefore it is necessary to ensure the adequacy of the stress distribution and check against different types of postulated failure modes. The proposed design is finally iterated until the most economical and reliable product is obtained. The functional requirements cover the geometrical design parameters such as size and shape, location of the penetrations, and so on. Some of these parameters may have to be fixed in collaboration with the overall design team, but in a majority of situations the pressure vessel designer acts freely on the basis of his or her experience.

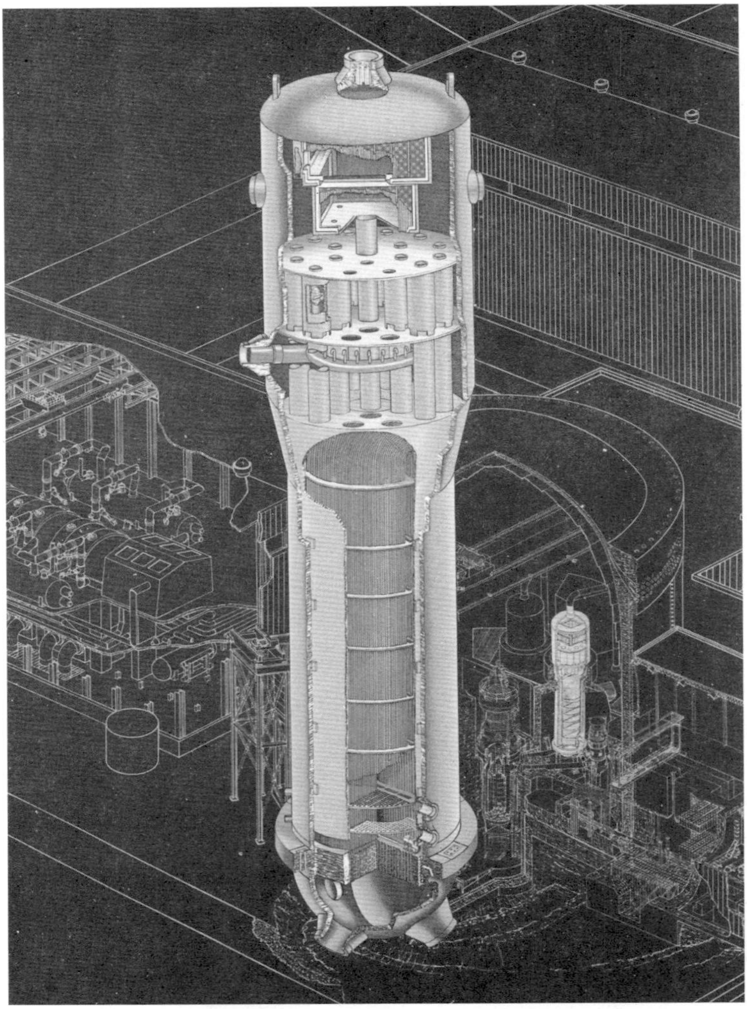

Figure 1.5 Pressurized water reactor (PWR) steam generator.
(Courtesy of Westinghouse Electric Company, Pittsburgh, PA.)

In the design of pressure vessels safety is the primary consideration, especially for nuclear reactor pressure vessels, due the potential impact of a possible severe accident. In general however, the design is a compromise between consideration of economics and safety. The possible risks of a given mode of failure and its consequences are balanced against the effort required for its prevention; the resulting design should achieve an adequate standard of safety at minimum cost.

Safety cannot be absolutely assured for two reasons. First, the actual form of loading during service may be more severe than was anticipated at

Figure 1.6 Pressurized water reactor (PWR) pressurizer.
(Courtesy of Westinghouse Electric Company, Pittsburgh, PA.)

the design stage: abnormal, unpredictable loads inevitably occur during the pressure vessel's lifetime. Second, our knowledge is seldom adequate to provide a qualified answer to the fracture of materials, state of stress under certain conditions, and so on.

It is true that although the fundamental mechanism of failure is not sufficiently understood, it is possible to establish preventive measures based on semiempirical methods. Following this line of thinking, the pressure vessels could be classified according to the severity of their operations since this will affect both the possibility of failure and its

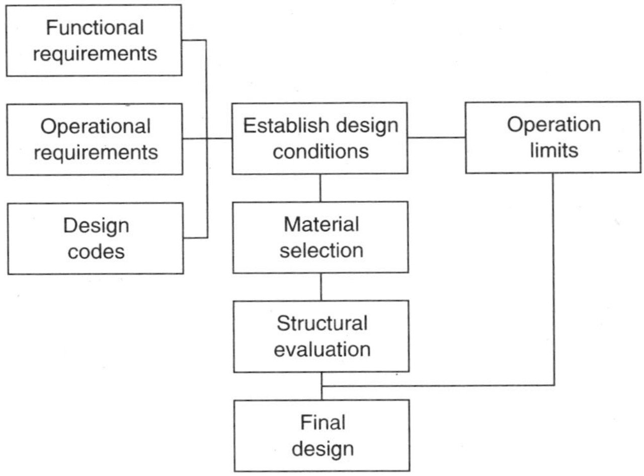

Figure 1.7 Design procedure.

consequences. These considerations lead to the classification of vessels ranging from nuclear reactor pressure vessels at one end to underground water tanks at the other. The design factor used in the ASME Boiler and Pressure Vessel Code[1] is intended to account for unknown factors associated with the design and construction of the equipment. The design formulas and the stress analysis methods are generally approximate and have built-in assumptions. Typically it is assumed that the material is homogeneous and isotropic. In the real world the material has flaws and discontinuities, which tend to deviate from this assumption.

In 1925, the rules for construction of power boilers was written using a design factor of 5 which was subsequently reduced to 4 in 1942, presumably to help conserve steel. In 1955, new processes in the petrochemical industry were requiring significant design pressures requiring wall thickness in vessels to be between 150 and 200 millimeters (mm). The ASME Pressure Vessel and Code Committee decided to form a task group with the allowable stresses based on a design factor of 3. The purpose was to reduce fabrication costs, with the implied assumptions that this could be applied to limited materials, with the addition of fracture toughness rules along with design rules for cyclic operation (fatigue) and that detailed stress analysis was used for most loading conditions. The committee felt that the nuclear code for pressure vessels would be easier to write than the code for pressure vessels used in petrochemical processes. This is because the nuclear pressure vessels only contained steam and water, and the maximum temperature was 800°F (427°C). Many nuclear plant design specifications identified that the design cycles for fatigue evaluation should be based on a 40-year life expectancy of the plant. The 40 years was based on nuclear plants being able to last twice as long as fossil plants (which usually lasted

20 years). The design for cyclic operation was based on the estimated cycles for 40 years, with the severity of the cycles based on estimated worst conditions. This method of design was a rough attempt to ensure freedom from fatigue cracking during the 40-year period. Using the code fatigue curves, a cumulative usage factor was calculated that was arbitrarily required to be equal to less than unity, which is based on the estimated number of cycles for the postulated 40-year period. The methodology has many conservative design factors in it, namely a factor of 2 for stress and a 20 for cycles.

1.2 Development of pressure vessel construction codes

Numerous boiler explosions took place through the late 1800s and early 1900s. This led to the enactment of the first code for construction of steam boilers by the Commonwealth of Massachusetts in 1907. This subsequently resulted in the development and publication of the ASME Boiler and Pressure Vessel Code in 1914, which sought to standardize the design, manufacturing, and inspection of boilers and pressure vessels. In 1921 the National Board of Boiler and Pressure Vessel Inspectors was organized to promote consistent inspection and testing. The publication of the section on locomotive boilers also appeared in 1921. The ASME and the ASTM (American Society for Testing and Materials) material specification merged in 1924. The first publication of Section VIII "Unfired Pressure Vessels," appeared in 1925. This document was referred to as one of a theoretical factor of safety of 5. The petroleum industry did not consider it to be adequate for their purposes and also desired better utilization of available materials. The year 1928 saw the advent of welded pressure vessels. For higher pressures the welded shells were made thicker than 70 mm. These required nondestructive examination (NDE) before service. In 1934, a joint API–ASME Committee published the first edition of an unfired pressure vessel code specifically for the petroleum industry. In 1952 these two separate codes merged into a single code – the ASME Unfired Pressure Vessel Code, Section VIII. The ASME Pressure Vessel Code, Section VIII Division 2: "Alternative Rules for Pressure Vessels," was published in 1968 and the original code became Section VIII Division 1: "Pressure Vessels."

A considerable boost was provided to the understanding of the basic behavior of pressure vessel components following the development of the nuclear power program in the U.S. and Europe in the late 1950s and early 1960s. Similar developments can be found in the British, French, German and Japanese codes, to name but a few. By 1960 the need for a code for pressure vessels for commercial nuclear plants became imperative. This resulted in publication of the 1963 Edition, Section III: "Nuclear Pressure Vessels." This was a design by analysis code with a theoretical safety factor of 3. After the publication of Section III: "Nuclear Pressure Vessels" in 1963, it was necessary to modify Section VIII for general pressure vessels. ASME

Code Section VIII Division 2: "Alternate Rules for Pressure Vessels" appeared as a result and provided a theoretical factor of safety of 3. In 1971, Section III: "Nuclear Power Components" were classified as (a) pumps, (b) valves, and (c) piping. The stress limits for emergency and faulted conditions were introduced. In addition, the addenda of 1971 added storage tanks. The addenda of summer 1972 introduced Appendix G on nonductile failure. The Appendix F on evaluation of faulted conditions was included in the addenda of winter 1972. The design of component supports and core support structures appeared in the addenda of winter 1973.

ASME Section III Division 1 is devoted entirely to nuclear power components and also contains the rules for the design of nuclear pumps and valves. The recognition of concrete reactor and containment vessels led to the publication of the Section II Division 2 code in 1975. Three subsections (NB, NC and ND) of ASME Section III Division 1 cover the design and construction of equipment of Classes 1, 2, and 3, respectively. The most stringent is Class 1, which requires design by analysis. Class 2 permits design by analysis as well as the use of formulas. Class 3 prescribes design by formula, and is equivalent to Section VIII Division 1. The designer evaluates the safety function of each pressure vessel and applies the appropriate code class. Design of supports for Section III Division 1 vessels are not prescribed in the ASME Code. Section III has a subsection NF, which prescribes the design of supports for Class 1, 2, and 3 pressure vessels. The addenda of winter 1976 changed the nomenclature of design, normal, upset, testing and faulted conditions to level A, B, C and D service conditions. In the 1982 addenda, the fatigue curves were extended to 10^{11} cycles. In the 1996 addenda, the design rules for high-temperature service were incorporated. In 1976, Division 3 was published which contained rules on transport of irradiated materials. The need for uniform rules for in-service inspection of nuclear power plants led to the issuance of the 1970 edition of Section XI: "Rules for In-service Inspection of Nuclear Plant Components."

The organization of the ASME Boiler and Pressure Vessel Code is as follows:

1. Section I: Power Boilers
2. Section II: Material Specification:
   i. Ferrous Material Specifications – Part A
   ii. Non-ferrous Material Specifications – Part B
   iii. Specifications for Welding Rods, Electrodes, and Filler Metals – Part C
   iv. Properties – Part D

3. Section III Subsection NCA: General Requirements for Division 1 and Division 2
   i. Section III Division 1:
      a. Subsection NA: General Requirements
      b. Subsection NB: Class 1 Components

c. Subsection NC: Class 2 Components
d. Subsection ND: Class 3 Components
e. Subsection NE: Class MC Components
f. Subsection NF: Component Supports
g. Subsection NG: Core Support Structures
h. Appendices: Code Case N-47 Class 1: Components in Elevated Temperature Service

ii. Section III, Division 2: Codes for Concrete Reactor Vessel and Containment

4. Section IV: Rules for Construction of Heating Boilers
5. Section V: Nondestructive Examinations
6. Section VI: Recommended Rules for the Care and Operation of Heating Boilers
7. Section VII: Recommended Guidelines for Care of Power Boilers
8. Section VIII
   i. Division 1: Pressure Vessels – Rules for Construction
   ii. Division 2: Pressure Vessels – Alternative Rules
9. Section IX: Welding and Brazing Qualifications
10. Section X: Fiberglass-Reinforced Plastic Pressure Vessels
11. Section XI: Rules for In-Service Inspection of Nuclear Power Plant Components

The rules for design, fabrication and inspection of pressure vessels are provided by codes that have been developed by industry and government in various countries and are indicated in Table 1.1. The design and construction codes all have established rules of safety governing design, fabrication and inspection of boilers, pressure vessels and nuclear components. These codes are intended to provide reasonable protection of life and property and also provide for margin for deterioration in service. Table 1.1 also includes the ASME Boiler and Pressure Vessel Code. Some of the significant features of the latest version of the ASME Code Section III are:

- Explicit consideration of thermal stress
- Recognition of fatigue as a possible mode of failure
- The use of plastic limit analysis
- Reliable prediction of ductile failure after some plastic action.

In addition there is a continuous attempt to understand all failure modes, and provide rational margins of safety against each type of failure. These margins are generally consistent with the consequence of the specific mode of failure.

A word or two about the impact of technological advances in pressure vessel design should be mentioned. The last three decades have seen great strides made in the improvement of digital computations. In the 1960s the use of computers began to make an impact on design and analysis of

Table 1.1 Design and Construction Codes for Pressure Vessels

Country	Code	Issuing authority
U.S.	ASME Boiler & Pressure Vessel Code	ASME
U.K.	BS 1515 Fusion Welded Pressure Vessels BS 5500 Unfired Fusion Welded Pressure Vessels	British Standard Institute
Germany	AD Merblatter	Arbeitsgemeinschaft Druckbehalter
Italy	ANCC	Associazione Nationale Per Il Controllo Peula Combustione
Netherlands	Regeis Voor Toestellen	Dienst voor het Stoomvezen
Sweden	Tryckkarls kommissionen	Swedish Pressure Vessel Commission
Australia	AS 1200:SAA Boiler Code AS 1210 Unfired Pressure Vessels	Standards Association of Australia
Belgium	IBN Construction Code for Pressure Vessels	Belgian Standards Institute
Japan	MITI Code	Ministry of International Trade and Industry
France	SNCT Construction Code for Unfired Pressure Vessels	Syndicat National de la Chaudronnerie et de la Tuyauterie Industrielle

pressure vessels. The rapid development of finite-element software has remarkably impacted the detailed design of pressure vessel components. These developments along with continuing increase in computing speed and storage capacity of the computer have really made the design process extremely quick and at the same time have led to very accurate design assessment. Initially in the early to mid-1970s, detailed finite-element analyses were generally performed for confirmatory analyses. Today these tasks are routinely accomplished in an interactive mode. The three-dimensional finite-element analysis programs using solid elements are rapidly replacing plate, shell, and two-dimensional programs for routine structural design analysis of pressure vessels. In addition the concepts of computer-aided design (CAD) and computer-aided manufacturing (CAM) are being integrated.

In spite of some of the most rigorous, well-conceived safety rules and procedures ever put together, boiler and pressure vessel accidents continue to occur. In 1980, for example, the National Board of Boiler and Pressure Vessel Inspectors reported 1972 boiler and pressure vessel accidents, 108 injuries and 22 deaths.[2] The pressure vessel explosions are of course rare nowadays and are often caused by incorrect operation or poorly monitored corrosion. Safety in boiler and pressure vessels can be achieved by:

- Proper design and construction
- Proper maintenance and inspection
- Proper operator performance and vessel operation.

The design and construction cures are dependent upon the formulation and adoption of good construction and installation codes and standards. Thus the ASME Pressure Vessel Code requires that all pressure vessels be designed for the most severe coincident pressure and temperature expected during the intended service. There can be no deviation from this requirement, even if the severe condition is short term and occurring only occasionally. Bush has presented statistics of pressure vessels and piping failures in the U.S., Germany and the UK.[3] He has concluded that a 99 percent confidence upper boundary for the probability of disruptive failure to be less than 1×10^{-5} per vessel year in the U.S. and Germany. According to his study, periodic inspection is believed to be a significant factor in enhancing pressure vessel reliability, and successful applications of ASME Boiler and Pressure Vessel Codes (Sections I and VIII) are responsible for the relatively low incidence of noncritical failures early in life.

Pierre and Baylac authored an international perspective of the design of pressure vessels in 1992.[4] They recommend that the governing authorities be vigilant by constantly monitoring accident statistics. They also insist that the authorities be prudent and maintain a flexible attitude in enforcing regulations.

References

1. American Society of Mechanical Engineers, Boiler and Pressure Vessel Code, ASME, New York.
2. Jagger, R., Causes of boiler and pressure vessel accidents, *Prof. Saf.*, 29, 39–42, 1984.
3. Bush, S.H., Statistics of pressure vessel and piping failures, *Am. Soc. Mech. Eng. J. Pressure Vessel Technol.*, 110, 225–233, 1988.
4. Pierre, D., and Baylac, G., French pressure vessel regulations within the European context, *J. Pressure Technol.*, 114, 486–488, 1992.

Additional Readings

Cepluch, R.J., The ASME Boiler and Pressure Vessel Code Committee – challenges: past and future, *ASME J. Pressure Vessel Technol.*, 109, 256–259, 1987.

Merrick, R.D., Design of pressure vessels and tanks to minimize corrosion, *Mater, Performance*, 29–37, 1987.

Nasman, G.D., Pressure vessels: total design specification, *Chem. Eng.*, 72–76, 1990.

Pai, D.H., Pressure vessel and piping technology: two decades of progress and future challenges, *J. Pressure Vessel Technol.*, 112, 319–322, 1990.

Windenberg, D.F., and Trilling, C., *Pressure Vessel and Piping Design: Collected Papers*, American Society of Mechanical Engineers, New York, 1960.

chapter two

Pressure vessel design philosophy

Contents

2.1 General overview

Engineering design is an activity to ensure fitness for service. Within the context of pressure vessel design, this primarily involves strength considerations. The "total design" is a topic with far-reaching ramifications. It might include aspects of fuel system design, reactor design, or thermal hydraulic design. In our subsequent discussions, the underlying philosophy, decisions and calculations related solely to the strength design are referred to the "pressure vessel design." For certain pressure vessels and related equipment, preliminary design may still be governed by heat transfer and fluid flow requirements. Although the aspect of thermal hydraulic design is intricately related to the structural design, especially for thermal transient loadings, we will not be discussing them in any detail. It will be assumed that the temperature distribution associated with a particular thermal transient has already been evaluated in a typical design application. However, in these cases the designer still has to consider how the desired configurations of the vessel are to be designed from a structural standpoint and how these designs will perform their intended service.

The role of engineering mechanics in the pressure vessel design process is to provide descriptions of the pressure vessel parts and materials in terms of mathematical models, which can be analyzed in closed form in a limited number of situations and mostly have to be solved numerically. Even the so-called simple models that can be solved in closed form might involve fairly complex mathematics. In a few isolated instances, intelligent applications of well-known principles have led to simplifying concepts. These concepts have generally eased the designer's task. However, in a majority of cases, especially when advanced materials and alloys are at a premium, there is a need to make the optimum use of the materials necessitating application of advanced structural analysis. As the complexity of the analysis increases, the aspect of interpretation of the results of the analysis becomes increasingly extensive. Furthermore, a large number of these models approximate the material behavior along with the extent of yielding. As we understand material behavior more and more, the uncertainties and omitted factors in design become more apparent. The improvement will continue as knowledge and cognizance of influencing design and material parameters increase and are put to engineering and economic use.

The safety demands within the nuclear industry have accelerated studies on pressure vessel material behavior and advanced the state of the art of stress analysis. For instance, the nuclear reactor, with its extremely large heavy section cover flanges and nozzle reinforcement operating under severe thermal transients in a neutron irradiation environment, has focused considerable attention on research in this area which has been directly responsible for improved materials, knowledge of their behavior in specific environments, and new stress analysis methods.

High-strength materials created by alloying elements, manufacturing processes, or heat treatments, are developed to satisfy economic or engineering demands such as reduced vessel thickness. They are continually being tested to establish design limits consistent with their higher strength and adapted to vessel design as experimental and fabrication knowledge justifies their use. There is no one perfect material for pressure vessels suitable for all environments, but material selection must match application and environment. This has become especially important in chemical reactors because of the embrittlement effects of gaseous absorption, and in nuclear reactors because of the irradiation damage from neutron bombardment.

Major improvements, extensions and developments in analytical and experimental stress analysis are permitting fuller utilization of material properties with confidence and justification. Many previously insoluble equations of elasticity are now being solved numerically. These together with experimental techniques are being used to study the structural discontinuities at nozzle openings, attachments, and so on. This is significant because 80 percent of all pressure vessel failures are caused by highly localized stresses associated with these "weak link" construction details. It is therefore apparent that the stress concentrations at vessel nozzle openings, attachments, and weldments are of prime importance, and methods for minimizing them through better designs and analyses are the keys to long pressure vessel life. Control of proper construction details results in a vessel of balanced design and maximum integrity.

In the area of pressure vessel design there are important roles played by the disciplines of structural mechanics as well as material science. As mentioned earlier, we try to provide a description of pressure vessel components in terms of mathematical models that are amenable to closed-form solutions, as well as numerical solutions. The development of computer methods (sometimes referred to as computer-aided design, or CAD) has had a profound impact on the stress and deflection analysis of pressure vessel components. Their use has been extended to include the evaluation criteria as well, by a suitable combination of postprocessing of the solutions and visual representation of numerical results. In a number of cases advanced software systems are dedicated to present animation that aids the visualization and subsequent appreciation of the analysis. A number of design and analysis codes have been developed that proceed from the conceptual design through the analysis, sometimes modeling the nonlinear geometric and material behavior. Results such as temperatures, deflections and stresses are routinely obtained, but the analysis often extends to further evaluations covering creep, fatigue, and fracture mechanics. With the advent of three-dimensional CAD software and their parametric, feature-driven automated design technology, it is now possible to ensure the integrity of designs by capturing changes anywhere in the product development process, and updating the model and all engineering deliverables automatically. Pressure vessel designs that once averaged 24 hours to finish are completed in about 2 hours. Such productivity gains translate into substantial savings in engineering labor associated with each new pressure vessel design.

The typical design of a pressure vessel component would entail looking at the geometry and manufacturing construction details, and subsequently at the loads experienced by the component. The load experienced by the vessel is related to factors such as design pressure, design temperature, and mechanical loads (due to dead weight and piping thermal expansion) along with the postulated transients (typically those due to temperature and pressure) that are anticipated during the life of the plant. These transients generally reflect the fluid temperature and pressure excursions of the mode of operation of the equipment. The type of fluid that will be contained in the pressure vessel of course is an important design parameter, especially if it is radioactive or toxic. Also included is the information on site location that would provide loads due to earthquake (seismic), and other postulated accident loads.

In assessing the structural integrity of the pressure vessel and associated equipment, an elastic analysis, an inelastic analysis (elastic–plastic or plastic) or a limit analysis may be invoked. The design philosophy then is to determine the stresses for the purpose of identifying the stress concentration, the proximity to the yield strength, or to determine the shakedown limit load. The stress concentration effects are then employed for detailed fatigue evaluation to assess structural integrity under cyclic loading. In some situations a crack growth analysis may be warranted, while in other situations, stability or buckling issues may be critical. For demonstrating adequacy for cyclic operation, the specific cycles and the associated loadings must be known *a priori*.

In this context, it is important for a pressure vessel designer to understand the nature of loading and the structural response to the loading. This generally decides what type of analysis needs to be performed, as well as what would be the magnitude of the allowable stresses or strains. Generally the loads acting on a structure can be classified as sustained, deformation controlled, or thermal. These three load types may be applied in a steady or a cyclic manner. The structure under the action of these loads may respond in a number of ways:[2]

- When the response is elastic, the structure is safe from collapse when the applied loading is steady. When the load is applied cyclically a failure due to fatigue is likely; this is termed *failure due to high cycle fatigue*.
- When the response is elastic in some regions of the structure and plastic in others, there is the potential to have an unacceptably large deformation produced by both sustained and deformation-controlled loads. Cyclic loads or cyclic temperature distributions can produce plastic deformations that alternate in tension and compression and cause fatigue failure, termed *low cycle fatigue*. Such distribution of loads could be of such a magnitude that it produces plastic deformations in some regions when initially applied, but upon removal these deformations become elastic, and subsequent loading results in predominantly elastic action. This is termed *shakedown*. Under cyclic loading fatigue failure is likely and because of elastic action, this would be termed as low cycle fatigue.
- When the sustained loading (due to bending or tension) is such that the entire cross-section becomes plastic, gross collapse of the structure takes place.
- Ratcheting is produced by a combination of a sustained extensional load and either a strain-controlled cyclic load or a cyclic temperature distribution that is alternately applied and removed. This produces cycling straining of the material which in turn produces incremental growth (cyclic) leading to what is called an incremental collapse. This can also lead to low cycle fatigue.
- Sustained loads in brittle materials or in ductile materials at low temperatures could result in brittle fracture, which is a form of structural collapse.

2.2 *Structural and material considerations*

The continued and prolonged use of pressure vessels for power generation, nuclear or chemical reactions, industrial processing, and storage requires them to withstand severe conditions of pressure, temperature, and other environments. Such environmental conditions include corrosion, neutron irradiation, hydrogen embrittlement, and so on. Pressure vessels are required to operate at a temperature range from as high as 600°C to as low as –20°C, with design pressures as high as 140 MPa. Some vessels are designed to carry noncorrosive fluids; while others are designed to withstand harsh corrosive and highly radioactive environments. The type of service, whether steady or cyclic, may also vary considerably. For each set of operating parameters, the pressure vessel material may be required to have certain properties. For example, operation at very low temperatures would require the use of materials with high notch toughness, while operation at high temperatures would require materials with high creep strength. Apart from the mechanical properties, considerations on manufacturability, commercial availability, as well as cost, has to be accounted for in the selection process.

The materials that are used in pressure vessel construction are:

- Steels
- Nonferrous materials such as aluminum and copper
- Specialty metals such as titanium and zirconium
- Nonmetallic materials, such as, plastic, composites and concrete
- Metallic and nonmetallic protective coatings

The mechanical properties that generally are of interest are:

- Yield strength
- Ultimate strength
- Reduction of area (a measure of ductility)
- Fracture toughness
- Resistance to corrosion

The failures that the pressure vessels are to be designed against are generally stress dependent. For this reason it becomes necessary to obtain the stress distribution in the pressure vessels. There is a need to evaluate the operating stresses due to the imposed conditions by analytical methods and sometimes by experimental means. Furthermore we also need to understand the significance of these stresses on the structural integrity of the pressure vessel by considering the material properties of the vessel. Developments in aerospace, nuclear, chemical, and petrochemical industries have put demands on pressure vessel materials to sustain thermal shock, dynamics, and cyclic operation (fatigue). Knowledge of the material behavior is necessary not only to ensure that the vessel can withstand the loading but also to make sure that the material has been chosen and utilized in an optimum manner.

The requirements that are imposed on the design of a pressure vessel by the mode of operation specified for the overall plant are divided into two groups. The first group includes those resulting from the operation at maintained loading either under maximum or normal conditions. For this group the operating pressure (internal and external) existing during the normal operation is required. The second group includes the transient conditions that exist during start-up or shutdown or during a general change in loading. For this group it is necessary to know the maximum maintained pressure that may be anticipated.

The fluid temperature is another fundamental requirement. The maximum and minimum values as well as the history of temperature variation need to be known. The material selection is dictated to some extent by this requirement. Further requirements might involve environmental characteristics such as corrosion, erosion, and irradiation.

Mechanical loads on the pressure vessel include those due to:

- Pressure
- Dead weight
- Seismic factors
- Piping

In addition, snow and wind loadings should be considered wherever applicable. Other loads due to various postulated accidents must also be considered.

Pressure vessels are designed for a postulated or expected design life. In addition the possibility of periodic inspections is of importance. Thus it is required to provide inspection ports in terms of handholes or manholes as necessary. The detailed description of the mode of operation, the definition of the rate of change of fluid temperatures as well as the number of occurrences of various transient events need to be specified.

The vessels need to be designed according to the severity of operation. For example, pressure vessels for nuclear applications have to be designed according to postulated accidents and associated possible risks of failure, including the release of radioactive materials. This is also the case for vessels with corrosive fluids at high pressure. The energy released in the event of a catastrophic failure is an important consideration in the design of vessels. These considerations lead to a classification of vessels varying from nuclear reactor vessels at one end of the scale to underground water tanks at the other. The designer uses his or her own discretion as to the position of the particular design in the scale of the severity.

The stress level is maintained below the allowable level, which is based on consideration of many failures; for example, plastic collapse, fatigue, brittle fracture, or buckling.

Stress analysis involves determining the relationship between the applied loads on the vessel and the associated response in terms of deflections, stresses, and strains.

When a bar is subjected to a tensile test we can obtain the stress–strain curve for the material. A common tensile test specimen is shown in Figure 2.1, and a typical stress–strain curve is shown in Figure 2.2.[3] The curve generally contains a linear portion depicting the material's elastic behavior, the modulus of elasticity being the ratio of stress to strain in this region. Beyond the linear portion, or beyond the elastic limit, the stress–strain characteristic is usually nonlinear, and the material is subject to nonrecoverable plastic deformation. The 0.2 percent offset yield strength refers to the stress which when removed leaves a permanent strain of 0.2 percent in the material. This is the practical definition of the yield strength of a material. When the specimen breaks, the nominal level of stress is the ultimate strength of the material.

The older design procedures of pressure vessels were based on sustained loading and on the concepts of the static strength of materials. These were mostly appropriate and adequate, because repetitive loadings were uncommon and parts were designed with ample factors of safety. In recent years, with the development and use of power machinery and equipment, inexplicable failures of ductile materials at stresses below the ultimate strength and sometimes even below the yield strength have taken place. These have been attributed to fatigue, since these failures tended to appear after a period of service. It has been established that the important factor is the repetition of stress rather than the duration of time at a particular stress level.

The modern view of the fatigue process is characterized by three main stages:

1. Fatigue crack initiation
2. Fatigue crack growth to a critical size
3. Failure of the net section.

The crack is generally believed to initiate at a surface flaw and to spread from this location during the stress cycling until the section is reduced sufficiently for an eventual tensile fracture to take place. Since fatigue

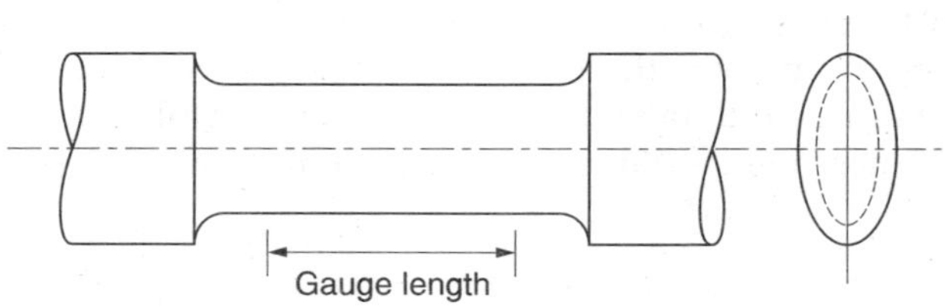

Figure 2.1 Tensile test specimen.

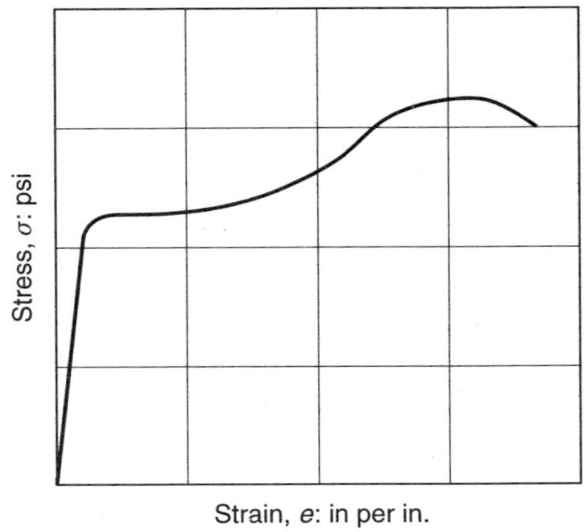

Figure 2.2 Typical stress–strain curve.

failure involves the combined effect of a number of small-scale events taking place over many stress cycles, it is fairly difficult to predict the fatigue life. Some aspects of fatigue, however, can be addressed in a semiempirical way. Cyclic testing may be performed for direct compression and tension, bending, torsion, and in some cases a combination of these factors. The simplest and most frequently used method is the R.R. Moore rotating reversed beam bending test. Here, the beam specimen is subjected to bending by a load applied at its center while being rotated at a constant speed, thus creating a completely reversed bending stress with each revolution. Data from such tests are termed *S–N* curves – the abscissa indicating the stress level and the ordinate representing the number of cycles to failure. A typical *S–N* curve for mild steel is shown in Figure 2.3. Initially the stress level *S* decreases with increase in the number of cycles *N*, then the curve is shown to approach asymptotically a constant stress value beyond which no further reduction in *S* takes place with increasing *N*. This is called the endurance limit of the material. This is not a universal property for all materials; only for some ferritic steels this endurance limit is realized between 10^6 to 10^7 cycles. For other materials *S* is seen to drop, albeit at a small rate, with the number of cycles.

Actual service conditions are often characterized by a number of cycles of stress of different magnitudes. One method of assessing this failure from repetitive stresses involves the concept of cumulative damage and posits that fatigue failure will take place when the cumulative damage (the summation of incremental damages) equals unity. This is represented as:

$$\sum_{1}^{m} \frac{n_i}{N_i} = 1 \qquad (2.1)$$

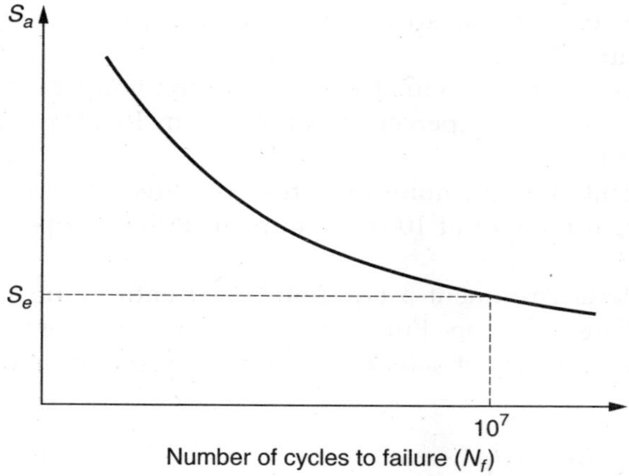

Figure 2.3 S–N curve.

where n_i is the number of cycles at stress level σ_i, and N_i is the number of cycles to failure at the same stress σ_i. The ratio n_i/N_i is the incremental damage or the cycle ratio, and it represents the fraction of the total life that each stress ratio uses up. If the sum of all the different stress cycles (m) is less than unity the vessel is presumed safe.

2.3 Factor of safety

The design equations in the various codes of construction always contain factors of safety. Realistically this factor is intended to account for the uncertainties in load, the dimensions, and the material properties. The approach taken in pressure vessel design, however, is to incorporate the types of material properties relevant to different modes of failure. These safety factors are strongly dependent on the modes of failure, as indicated in the design equations. The safety factors are generally applied to the pressure vessel materials so that significant assurances exist that the component can safely perform in the operating environment. Because of the complexity and the multiplicity of demands placed on the material of construction, the allowable stresses (hence the safety factors) are not based on a single material property, but on a combination of a number of properties. These properties could be the tensile strength, the yield strength, elongation, and so on. For example, the American Society of Mechanical Engineers (ASME) Boiler and Pressure Vessel Code, Section VIII Division 1: "Rules for the Construction of Pressure Vessels," used for establishing allowable stress values, advocates using the lesser of the following:

- 25 percent of the specified minimum tensile strength at room temperature
- 25 percent of the tensile strength at design temperature

- 62.5 percent of the specified minimum yield strength at room temperature
- 62.5 percent of the yield strength at design temperature
- Stress to produce 1 percent creep strain in 100,000 hours at design temperature
- 80 percent of the minimum stress required to produce material rupture, at the end of 100,000 hours at design temperature

For a simple environment a criterion based entirely on yield strength seems appropriate, therefore European pressure vessel construction codes typically employ a factor of safety of 1.5 for the yield strength.

2.4 Design by rule

By following design-by-rule methods, the designer simply follows the rules laid out in the procedures for components such as nozzles, heads, flanges, and so on. This procedure has the great advantage of simplicity and consistency but has several limitations. For example, there are cases when the loadings and geometries are such that the procedure cannot be applied effectively. Some of the rules are based on elastic stress analysis with some limitations on maximum stress. Some are based on shakedown concepts without specifically considering stress ranges, while others are based on limit load concepts with suitable shape factors. Design-by-rule methods were used in earlier ASME design codes (Sections I and VIII).

Generally speaking, design-by-rule methods of design are based on experience and tests. This process requires the determination of design loads, the choice of a design formula and the selection of an appropriate stress allowable for the material used. The procedure provides the information on required vessel wall thickness as well as the rules of fabrication and details of construction. These rules do not typically address thermal stresses and fatigue. The fatigue issues are considered covered by the factors of safety.

2.5 Design by analysis

This philosophy originated in the 1960s and was motivated by the sophisticated design work performed in the nuclear industry at the time. It effectively integrates design and stress analysis efforts and recognizes that different stress states have different degrees of importance. Furthermore, this process accounts for most failure modes and provides rational margins of safety against each mode of failure. The process involves detailed evaluation of actual stress including thermal stresses and fatigue. This design approach provides a rational safety margin (not unduly excessive) based on the actual stress profile and optimizes design to conserve material, leading to consistent reliability and safety. This philosophy is appropriate for pressure vessels involving cyclic operation and requiring superior reliability and safety, and is suitable for pressure vessels for which periodic inspection is deemed difficult (e.g., nuclear vessels). This viewpoint was first incorporated into the ASME Boiler and Pressure Vessel Code Section III and Section VIII Division 2 in 1968.[1]

References

1. American Society of Mechanical Engineers, Boiler and Pressure Vessel Code, ASME, New York.
2. Burgreen, D., *Design Methods for Power Plant Structures*, C.P. Press, New York, 1975.
3. Harvey, J.F., *Theory and Design of Pressure Vessels*, Van Nostrand Reinhold, New York, 1985.

chapter three

Structural design criteria

Contents

3.1 Modes of failure

Two basic modes of failure are assumed for the design of pressure vessels. These are: (a) elastic failure, governed by the theory of elasticity; and (b) plastic failure, governed by the theory of plasticity. Except for thick-walled pressure vessels, elastic failure is assumed. When the material is stretched beyond the elastic limit, excessive plastic deformation or rupture is expected. The relevant material properties are the yield strength and ultimate strength. In real vessels we have a multiaxial stress situation, where the failure is not governed by the individual components of stress but by some combination of all stress components.

3.2 Theories of failure

The most commonly used theories of failure are:

- Maximum principal stress theory
- Maximum shear stress theory

- Maximum distortion energy theory

According to the maximum principal stress theory, failure occurs when one of the three principal stresses reaches a stress value of elastic limit as determined from a uniaxial tension test. This theory is meaningful for brittle fracture situations.

According to the maximum shear stress theory, the maximum shear equals the shear stress at the elastic limit as determined from the uniaxial tension test. Here the maximum shear stress is one half the difference between the largest (say σ_1) and the smallest (say σ_3) principal stresses. This is also known as the Tresca criterion, which states that yielding takes place when

$$\frac{(\sigma_1 - \sigma_3)}{2} = \pm\frac{\sigma_\mathrm{y}}{2} \tag{3.1}$$

The distortion energy theory considers failure to have occurred when the distortion energy accumulated in the component under stress reaches the elastic limit as determined by the distortion energy in a uniaxial tension test. This is also known as the von Mises criterion, which states that yielding will take place when

$$\frac{1}{\sqrt{2}}\left[(\sigma_1 - \sigma_2)^2 + (\sigma_2 - \sigma_3)^2 + (\sigma_3 - \sigma_1)^2\right] = \pm\sigma_y \tag{3.2}$$

To understand the essential differences between the Tresca and von Mises criteria let us consider the simplified case of a biaxial stress state, where we assume that the principal stress, σ_3 is zero.

Let us first consider the case of Tresca criterion. We further assume that σ_1 and σ_2 have the same sign. Then, following Eq. (3.1), we have

$$|\sigma_1 - \sigma_3| = \sigma_\mathrm{y} \tag{3.3a}$$

or

$$|\sigma_2 - \sigma_3| = \sigma_\mathrm{y} \tag{3.3b}$$

This gives

$$\sigma_1 = \sigma_\mathrm{y};\quad \sigma_1 = -\sigma_\mathrm{y};\quad \sigma_2 = \sigma_\mathrm{y};\quad \sigma_2 = -\sigma_\mathrm{y} \tag{3.4}$$

Next we that σ_1 and σ_2 are of the opposite sign. The yielding will then take place when

$$|\sigma_1 - \sigma_2| = \sigma_\mathrm{y} \tag{3.5}$$

This implies that

$$\sigma_1 - \sigma_2 = \sigma_y \tag{3.6a}$$

or

$$\sigma_2 - \sigma_1 = \sigma_y \tag{3.6b}$$

If Eqs. (3.4) and (3.6) are plotted with σ_1 as abscissa and σ_2 as the ordinate, then we get six straight lines (shown as dashed hexagon in Figure 3.1). The values of σ_1 and σ_2 falling on the hexagon and outside would cause yielding. We have of course assumed that the material yield strength is equal in magnitude when in tension or in compression.

Next we consider the von Mises criterion. With the assumption that $\sigma_3 = 0$, Eq. (3.2) gives

$$\sigma_1^2 - \sigma_1\sigma_2 + \sigma_2^2 = \sigma_y^2 \tag{3.7}$$

This equation is plotted in the $\sigma_1 - \sigma_2$ plot as shown in the solid lines (forming an ellipse) in Figure 3.1. According to the von Mises criterion, the points falling on or outside of the ellipse would cause yielding.

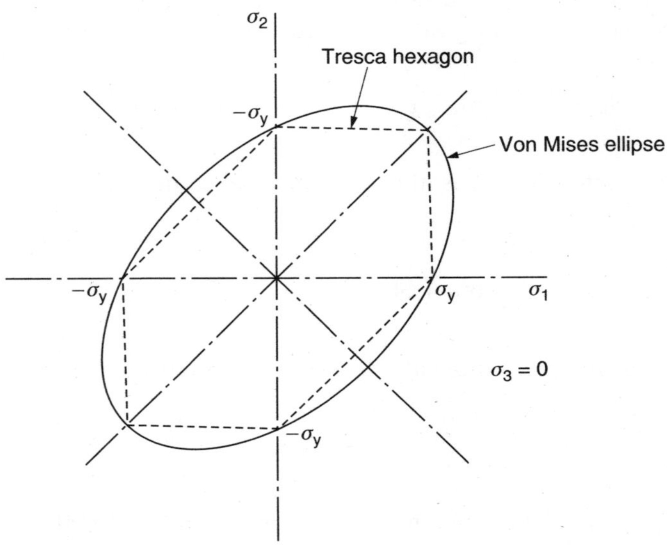

Figure 3.1 Tresca and von Mises theories of failure.

3.3 Theories of failure used in ASME Boiler and Pressure Vessel Code

Two basic theories of failure are used in the American Society of Mechanical Engineers (ASME) Boiler and Pressure Vessel Code, Section I, Section IV, Section III Division 1 (Subsections NC, ND, and NE), and Section VIII Division 1 use the maximum principal stress theory. Section III Division 1 (Subsection NB and the optional part of NC) and Section VIII Division 2 use the maximum shear stress theory or the Tresca criterion. The maximum principal stress theory (sometimes called Rankine theory) is appropriate for materials such as cast iron at room temperature, and for mild steels at temperatures below the nil ductility transition (NDT) temperature (discussed in Section 3.7). Although this theory is used in some design codes (as mentioned previously) the reason is that of simplicity, in that it reduces the amount of analysis, although often necessitating large factors of safety.

It is generally agreed that the von Mises criterion is better suited for common pressure vessels, the ASME Code chose to use the Tresca criterion as a framework for the design by analysis procedure for two reasons: (a) it is more conservative, and (b) it is considered easier to apply. However, now that computers are used for the calculations, the von Mises expression is a continuous function and is easily adapted for calculations, whereas the Tresca expression is discontinuous (as can be seen from Figure 3.1).

In order to avoid dividing both the calculated and the yield stress by two, the ASME Code defines new terms called stress intensity, and stress difference. The stress differences (S_{ij}) are simply the algebraic differences of the principal stresses, σ_1, σ_2, and σ_3, so that

$$S_{1,2} = \sigma_1 - \sigma_2, S_{2,3} = \sigma_2 - \sigma_3, S_{3,1} = \sigma_3 - \sigma_1 \tag{3.8}$$

The stress intensity, S, is the maximum absolute value of the stress difference

$$S = \max(|S_{1,2}|, \quad |S_{2,3}|, \quad |S_{3,1}|) \tag{3.9}$$

In terms of the stress intensity, S, Tresca criterion then reduces to

$$S = \sigma_y \tag{3.10}$$

Throughout the design by analysis procedure in the ASME Code stress intensities are used.

3.4 Allowable stress limits in the ASME Boiler and Pressure Vessel Code

The overall objective in determining the allowable stress limits is to ensure that a pressure vessel does not fail within its established design life. The modes that are most likely to cause a failure, as identified by the ASME Code, are as follows:[2]

- Excessive elastic deformation including elastic instability
- Excessive plastic deformation
- Brittle fracture
- Stress rupture or creep deformation (inelastic)
- Plastic instability and incremental collapse
- High strain and low cycle fatigue
- Stress corrosion
- Corrosion fatigue

The first failure mode, namely that of excessive elastic deformation, is generally related to functional requirements. The aspect of elastic instability deals with the propensity of buckling in thin shells. The aspect of excessive plastic deformation could lead to complete collapse as outlined in the previous chapter. This failure mode requires that the analysis be addressed from the standpoint of bursting and gross distortion from a single load application. The failure mode associated with brittle fracture is related to the fracture toughness and is addressed later in this chapter. The failure mode associated with stress rupture or creep is appropriate for pressure vessels operating at high temperatures and as such will not be discussed here. The failure associate with plastic instability and incremental collapse was identified in the previous chapter as ratcheting causing progressive growth due to cyclic load application and should be addressed at the analysis stage. The high strain and low cycle fatigue is an important consideration for cyclic thermal loads. The crack initiation from fatigue damage should be addressed in the analysis. The failure modes associated with stress corrosion and corrosion fatigue are related to the environmental considerations as well as mode of operation.

The allowable stress limits in the ASME Code are established on two modes of failure and are characterized as:

- Avoidance of gross distortion or bursting
- Avoidance of ratcheting

In order for sustained loads to produce collapse in a structure, it is necessary that the loads produce full plasticity over the cross-section bearing the load, leading to what is commonly termed as the "plastic hinge." The stresses they produce are designated primary stresses. The set of primary mean stresses (or primary membrane stresses), P_m, represent the sustained load acting on the structure divided by the cross-sectional area resisting the load. In fact P_m is the stress intensity derived from the stress distribution and as such is the difference between the largest and the smallest of the principal stresses. P_m determines the susceptibility of the structure to fail by plastic collapse. In order to avoid gross distortion it is necessary to avoid a significant portion of the wall of the vessel from becoming fully plastic. For an elastic–perfectly plastic stress strain law (Figure 3.2) such a vessel would be fully plastic when the membrane stress reaches the yield stress. A safety factor of 1.5 is provided to avoid this situation (see Figure 3.3 for the design limit for P_m/S_y). The allowable design stress (primary membrane) is therefore limited to a stress limit typically two-thirds of the yield (referred to as material allowable S_m).

Large bending moments acting over the full cross-section can also produce structural collapse. The set of bending stresses generated by sustained bending moments are termed primary bending stresses, P_b, and at any particular point in the structure, being the stress intensities, they represent the differences between the largest and the smallest values of the principal stresses. The mode of collapse is bending, as opposed to extension, and the collapse will take place only when there is complete plastic yielding of the net cross-section. The pattern of plasticity in this plastic hinge so formed, consists of part of the cross-section becoming plastic in tension and the remainder of the section becoming plastic in compression.

When there are both direct (membrane) as well as bending stresses, the avoidance of gross distortion or bursting in a vessel is treated in the same way as direct and bending stresses in a rectangular beam. If such a beam is loaded in bending, collapse does not occur until the load has been increased by a factor known as the "shape factor" of the cross-section when a plastic hinge is formed. The shape factor of a rectangular section in bending is 1.5.

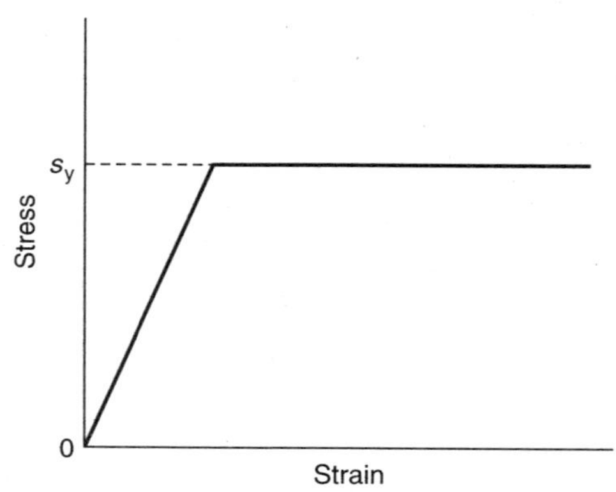

Figure 3.2 Strain–strain characteristics for an elastic–perfectly plastic material.

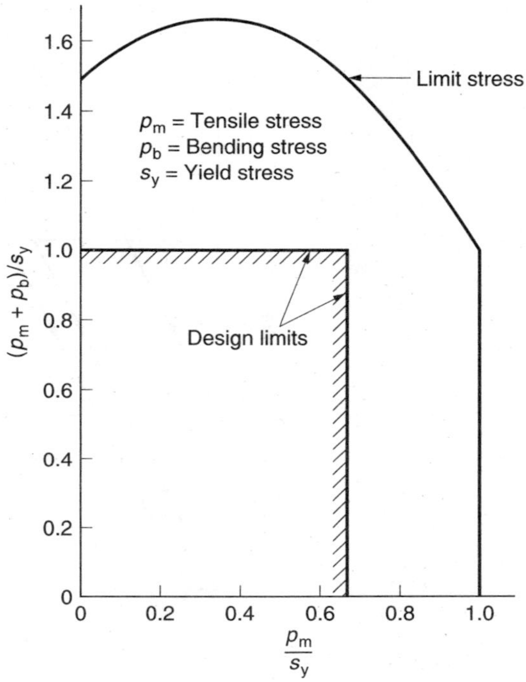

Figure 3.3 Membrane plus bending versus membrane stress for a rectangular beam.

When the primary stress in a rectangular section consists of a combination of bending and axial tension, the value of the limit load depends on the ratio between the tensile and bending loads. Figure 3.3 shows the value of the maximum calculated stress at the outer fiber of a rectangular section required to produce a plastic hinge plotted against the average tensile stress across the section, with both values expressed as multiples of the yield stress S_y. When the average tensile stress P_m is zero, the failure stress for bending is 1.5 S_y. The ASME Code limits the combination of the membrane and bending to the yield stress S_y. It can be seen from Figure 3.3 that there are variable margins depending on the particular combination of stresses, but it was decided to keep the design limits simple.

The repeated plastic straining or ratcheting is sometimes termed incremental collapse. If a structure is repeatedly loaded to progressively higher levels, one can imagine that at some highly stressed region a stage will be reached when the plastic strain will accumulate during each cycle of load, a situation that must be avoided. However, some initial plastic deformation is judged permissible during the first few cycles of load provided the structure shakes down to elastic behavior for subsequent loading cycles. Consider, for example, the outer fiber of a beam strained in tension to a value ε_1, somewhat beyond the yield strain as shown in Figure

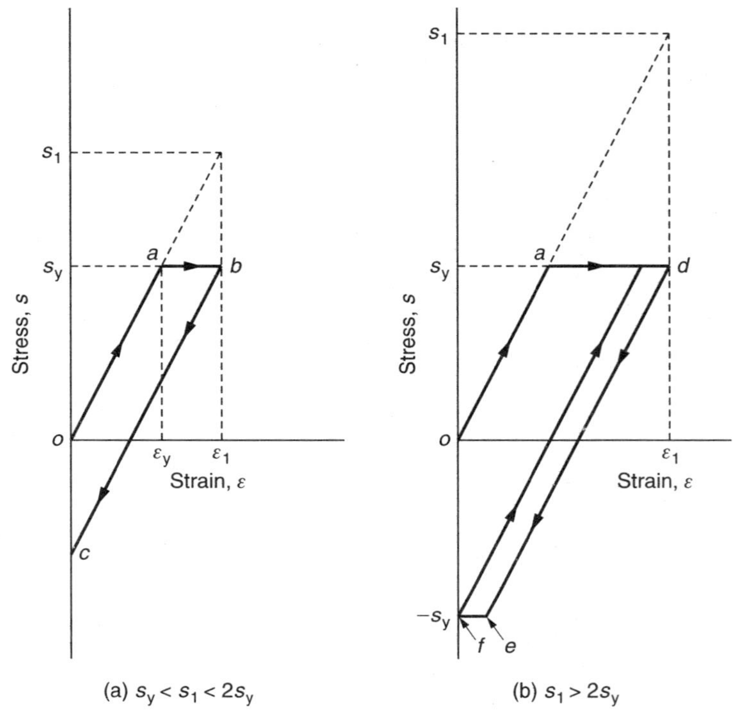

Figure 3.4 Ratcheting behavior.

3.4(a) by the path *OAB*. The calculated elastic stress would be $S = S_1 = E\varepsilon_1$. When the beam is returned to its undeformed position *O*, the outer fiber has a residual compressive stress of magnitude $S_1 - S_y$. On any subsequent loading, the residual compression must be removed before the stress goes into tension and thus the elastic stress range has been increased by the quantity $S_1 - S_y$. If $S_1 = 2S_y$, the elastic range becomes $2S_y$, but if $S_1 > 2S_y$, the fiber yields in compression, as shown by the line EF in Figure 3.4(b) and all subsequent cycles produce plastic strain. Therefore the limit of $2S_y$ could be regarded as a threshold beyond which some plasticity action would progress.

3.5 Service limits

The loading conditions that are generally considered for the design of pressure vessels include pressure, dead weight, piping reaction, seismic, thermal expansion and loadings due to wind and snow. The ASME Boiler and Pressure Vessel Code delineates the various loads in terms of the following conditions:

1. Design
2. Testing
3. Level A
4. Level B
5. Level C
6. Level D

Test conditions refer to the hydrostatic tests that are performed on the pressure vessel during its operating life. Level A service limits correspond to those of normal operating conditions. Level B service limits are sometimes referred to as "upset" conditions, and are those for which the component must withstand without sustaining damage requiring repair. Typically this includes the operating basis earthquake (OBE) and thermal transients for which the power level changes are on the order of 10 to 20 percent. Level C service limits constitute the emergency conditions in which large deformations in the area of discontinuity are created. Level D service limits are so called faulted conditions, for which gross deformation with a loss of dimensional stability is permitted. The component may require repair or removal. Examples are safe shutdown earthquake (SSE), pipe break or a combination of such events.

Specifically for the ASME Code, the primary membrane stress intensity, P_m, and the combined membrane plus bending stress intensity, $P_m + P_b$, (also the local membrane plus bending stress intensity, $P_L + P_b$ in some cases) for the various loading conditions are shown below.

1. Design condition:

$$\begin{aligned} P_m &\le S_m \\ P_m + P_b &\le 1.5S_m \end{aligned} \tag{3.11}$$

2. Testing condition:

$$\begin{aligned} &P_m \le 0.9S_y \\ &P_m + P_b \le 1.35S_y, \text{ for } P_m \le 0.67S_y \\ &P_m + P_b \le (2.15S_y - 1.2P_m) \text{ for } 0.67S_y \le P_m \le 0.9S_y \end{aligned} \tag{3.12}$$

3. Level C condition (emergency):

$$\begin{aligned} &P_m \le S_y \\ &P_L + P_b \le 1.5S_y, \text{ for } P_m \le 0.67S_y \\ &P_L + P_b \le (2.5S_y - 1.5P_L) \text{ for } P_L > 0.67S_y \end{aligned} \tag{3.13}$$

4. Level D condition (faulted):

$$\begin{aligned} P_m &\leq \text{lesser of } 0.7S_u \text{ and } 2.4S_m \\ P_m + P_b &\leq \text{lesser of } 1.05S_u \text{ and } 3.6S_m \end{aligned} \tag{3.14}$$

In the following chapter, the above limits have been critically appraised by introducing the shape factor of the cross-section. The above limits are strictly applicable for rectangular cross-sections. The new limits have been proposed and discussed.

3.6 Design for cyclic loading

Due to loads that are applied in a cyclic fashion the material can fail by fatigue when sufficient cycles of loading are applied. The number of cycles that will cause fatigue failure depends on the magnitude of strain that is incurred during each cycle of loading. Fatigue data are generally obtained at room temperature and plotted in the form of nominal stress amplitude (one half of stress range) versus number of cycles to failure. The stress range is obtained by multiplying the strain range from the fatigue test by the modulus of elasticity. The endurance limit is defined as the cyclic stress amplitude, which will not cause fatigue failure regardless of the number of applied cycles of stress. However, for pressure vessels sometimes the endurance limit and one-million cycle fatigue limit are used interchangeably. Pressure vessel codes commonly use a factor of safety of 2 on the fatigue stress and a safety factor of 20 on fatigue life (number of cycles to failure). The design for cyclic loading is performed to check whether a pressure vessel designed statically will not fail due to multiple stress cycling. The process entails:

1. Identifying design details which introduce stress concentrations and therefore potential sites for fatigue failure
2. Identifying cyclic (or repeated) stresses experienced during service
3. Using appropriate *S–N* curves and deducing design life.

The concept of cumulative damage factor is a simple yet reliable method to determine the factor of safety against fatigue failure. If N_i denotes the allowable number of cycles corresponding to a stress range S_i, then the usage factor U_i at the material point due to n_i applied number of cycles of stress range S_i is

$$U_i = \frac{n_i}{N_i} \tag{3.15}$$

If the material is subjected to m different cycles of frequency n_i and corresponding to stress ranges S_i ($I = 1, 2, \ldots m$), then the cumulative damage factor, U, is given by

$$U = \sum_{i=1}^{m} U_i = \sum_{i-1}^{m} \frac{n_i}{N_i} \tag{3.16}$$

Safety from fatigue failure requires

$$U \leq 1 \tag{3.17}$$

The ASME design fatigue curves are based on strain controlled data in which the best fit curves are constructed by a factor of 2 on stress or a factor of 20 on cycles to account for environment, size effect, and data scatter.

3.7 *Protection against fracture*

Pressure vessel materials are primarily steels, and the main point of concern is the effect of temperature on the fracture toughness of steel. Steels are generally ductile, but their resistance to brittle fracture diminishes as the temperature is lowered. The lower limit of the operating temperature is therefore determined by the transition point at which there is a change from ductile to brittle fracture. The value of the stress at fracture under those situations can be considerably lower than the yield strength. The fracture properties including the transition temperature depend on the composition, heat treatment, prior cold work, and the size of the flaws that may be present. As the carbon content is increased from 0.1 to 0.8 percent, the NDT (nil ductility transition) temperature increases from –45°C to +50°C. Small amounts of manganese or niobium can produce large decrease in transition temperature. The four design criteria for mild steels can be summarized as follows:

1. NDT design criterion: The maximum principal stress should not exceed 34.5 MPa, to assure fracture arrest at temperatures below NDT temperature.
2. NDT +17°C design criterion: The temperature of operation must be maintained above an NDT of +17°C, to assure that brittle fracture will not take place at stress levels up to one half the yield strength.
3. NDT +33°C design criterion: The temperature of operation must be maintained above an NDT of +33 °C, to assure that brittle fracture will not take place at stress levels up to the yield strength.
4. NDT +67°C design criterion: The temperature of operation must be maintained above an NDT of +67°C, to assure that brittle fracture will not take place at any stress level.

The margin of safety from brittle fracture is therefore dependent on the stress level as well as the expected minimum temperature of operation. Some design codes use a single margin of safety criterion based on energy absorption in a Charpy test conducted at the minimum expected temperature of operation.[1]

References

1. Burgreen, D., *Design Methods for Power Plant Structures*, C. P. Press, 1975.
2. Anon., Criteria of the ASME Boiler and Pressure Vessel Code for Design by Analysis in Sections III and VIII, Division 2, American Society of Mechanical Engineers, New York.

Problems

1. The in-plane normal stresses in a flat plate are 10 MPa and 60 MPa and the shear stress is 30 MPa. Find the stress intensity and the von Mises equivalent stress. What is the factor of safety corresponding to (a) Tresca criterion, and (b) von Mises criterion if the material yield strength is 150 MPa?
2. The in-plane stresses in a flat plate are –50 MPa and –150 MPa on two perpendicular planes and a shear stress of 40 MPa on those planes. Compute the maximum shear stress, the stress intensity and the von Mises equivalent stress. What is the factor of safety corresponding to (a) Tresca criterion, and (b) von Mises criterion if the material yield strength is 200 MPa?
3. The hoop stress in a cylindrical shell with closed ends is pR/t and the longitudinal stress is $pR/(2t)$, where p is the internal pressure, R the mean radius and t the thickness. If the shell is of diameter 0.5 m and a thickness of 12.5 mm, and is subjected to an internal pressure of 7 MPa, determine the maximum shear stress, the stress intensity and the von Mises equivalent stress. What is the factor of safety corresponding to (a) Tresca criterion, and (b) von Mises criterion if the material yield strength is 160 MPa?
4. A carbon steel pressure vessel is subjected to 1000 pressure cycles at an alternating stress of 300 MPa. At this alternating stress the number of cycles to failure is 7000 from the design fatigue curve. Subsequently the vessel is subjected to 400 temperature cycles at an alternating stress of 700 MPa for which the number of cycles to failure is 600 from the fatigue curve. Is the vessel adequate for the given cyclic loading?

chapter four

Stress categories and stress limits

Contents

4.1 Introduction

First of all we need to define the term stress. Stress is a tensor quantity (neither a vector nor a scalar) that depends on the direction of applied load as well as on the plane it acts. Generally speaking, at a given plane there are both normal and shear stresses. However, there are planes within a structural component (that is being subjected to mechanical or thermal loads) that contain no shear stress. Such planes are called principal planes and the directions normal to those planes are called principal directions. The normal stresses (only stresses in those planes) are called principal stresses. For a general three-dimensional stress state there are always three principal planes along which the principal stresses act. In mathematical terms we can say that the problem of principal stresses is an eigenvalue problem, with the magnitudes of the principal stresses being the eigenvalues and their directions (normal to the planes on which they act) being the eigenvectors. Principal stress calculations form an essential activity for a general stress analysis problem.

4.2 Stress intensity

Let us indicate the principal stresses by σ_1, σ_2, and σ_3. Then we define the stress differences by:

$$S_{1,2} = \sigma_1 - \sigma_2; \quad S_{2,3} = \sigma_2 - \sigma_3; \quad S_{1,3} = \sigma_1 - \sigma_3 \tag{4.1}$$

The stress intensity, *SI*, is then the largest absolute value of the stress differences, or in other words

$$SI = \max\left[|S_{1,2}|, |S_{2,3}|, |S_{1,3}|\right] \tag{4.2}$$

The computed stress intensity is then compared with the material allowables taking into consideration the nature of the loading. The material allowables are based on yield and ultimate strength of the material with an implied factor of safety.

Within the context of pressure vessel design codes, the comparison of the allowable strength of the material is always done with respect to the stress intensities. This puts the comparison in terms of the appropriate failure theory either the maximum shear stress theory (Tresca criterion) or the maximum distortion energy theory (von Mises criterion). These failure theories have been discussed in some detail in Chapter 3.

4.3 Categorization of stresses

Stresses are generally characterized as (a) primary stress, (b) secondary stress, or (c) peak stress. In the following discussion, the primary stresses will be denoted by *P*, the secondary stress by *Q* and the peak stress by *F*. These nomenclatures also apply to the ASME Boiler and Pressure Vessel Code.[1] We will now define each of the three categories of stress.

4.3.1 Primary stress

Primary stress is any normal stress or a shear stress developed by the imposed loading which is necessary to satisfy equilibrium between external and internal loads. These stresses are not self-limiting. If primary stresses are increased such that yielding through net section occurs, subsequent increase in primary stress would be through strain hardening until failure or gross distortion occurs. Generally primary stresses result from an applied mechanical load, such as a pressure load. The concept of equilibrium is based on a monotonic load and a lower bound limit load (for a discussion on limit load and the lower bound theorem see Appendix C). When the limit load is exceeded, gross deformation takes place, hence the qualification "not self-limiting."

A further elaboration of primary stresses is provided in a definition by Pastor and Hechmer where they state:

> Primary stresses are those that can cause ductile rupture or a complete loss of load-carrying capability due to plastic collapse of the structure upon a single application of load. The purpose of the Code limits on primary stress is to prevent gross plastic deformation and to provide a nominal factor of safety on the ductile burst pressure.[2]

Primary stresses are further divided into three types: general primary membrane (P_m), local primary membrane (P_L), and primary bending (P_b). Quite often the concepts of general primary membrane stress and local primary membrane stress are used interchangeably; the local primary membrane stress representing a general primary membrane stress along a local structural discontinuity. The rigorous definition of the general primary membrane stress is the average primary stress across a solid section produced by mechanical loads, and excludes discontinuities and concentrations. The local primary membrane stress is defined as also the average stress across any solid section, but includes discontinuities. However, the general primary membrane stress is one that is so distributed in the structure that no redistribution of load occurs as a result of yielding. The failure mode associated with the general primary membrane stress and the local primary membrane stress are meant to be different; the general primary membrane stress leads to gross distortion with no redistribution, and the local primary membrane stress to excessive plastic deformation with redistribution of load.

The primary bending is the component of primary stress proportional to the distance from the centroid of the solid section, and is produced by mechanical loads. This definition excludes discontinuities and concentration. The concept of bending stress is akin to the situation of beam bending, with a neutral axis along the center line with regions of tension and compression. The membrane stress is the component having a constant value through the section and represents an average value.

4.3.2 Secondary stress

Secondary stress originates through the self-constraint of a structure. This must satisfy the imposed strain or displacement (continuity requirement) as opposed to being in equilibrium with the external load. Secondary stresses are self-limiting or self-equilibrating. The discontinuity conditions or thermal expansions are satisfied by local yielding and minor distortions. The major characteristic of the secondary stress is that it is a strain-controlled condition. Secondary stresses occur at structural discontinuities and can be caused by mechanical load or differential thermal expansion. The local stress concentrations are not considered for secondary stresses. There is no need for further dividing the secondary stress into membrane and bending categories. In terms of secondary stress we imply secondary membrane and bending in combination.

4.3.3 Peak stress

Peak stress is the highest stress in a region produced by a concentration (such as a notch or weld discontinuity) or by certain thermal stresses. Peak stresses do not cause significant distortion but may cause fatigue failure. Some examples of peak stresses include thermal stresses in a bimetallic interface, thermal shock stresses (or stresses due to rapid change in the temperature of the contained fluid), and stresses at a local structural discontinuity.

Within the context of local primary membrane stress, P_L, as well as secondary stress, Q, the discontinuity effects need not be elaborated. The structural discontinuity can be either gross or local. Gross structural discontinuity is a region where a source of stress and strain intensification affects a relatively large portion of the structure and has a significant effect on the overall stress or strain pattern. Some of the examples are head-to-shell and flange-to-shell junctions, nozzles, and junctions between shells of different diameters or thicknesses.

Local structural discontinuity is a region where a source of stress or strain intensification affects a relatively small volume of material and does not have a significant effect on the overall stress or strain pattern or on the structure as a whole.

The stress classifications for various parts of a pressure vessel are indicated in Table 4.1 and are reproduced from the American Society of Mechanical Engineers (ASME) Boiler and Pressure Vessel Code Sections III and VIII.[1] It can be observed that the membrane stress is considered primary for mechanical loads. For a number of geometries and loading situations, the bending stress is considered secondary. The bending stress is considered primary when the net section experiences the applied bending moment.

4.4 Stress limits

The allowable stresses (or more correctly the stress intensities) in pressure vessel design codes such as the ASME Boiler and Pressure Vessel Code are not expressed in terms of the yield strength or the ultimate strength but instead as multiples of tabulated design value called the design stress intensity (denoted for example as S_m). This value is typically two-thirds of the yield strength of the material or for other cases one-third of the ultimate

Table 4.1 Classification of stresses

Vessel part	Location	Origin of stress	Type of stress	Classification
Cylindrical or spherical shell	Remote from discontinuities	Internal pressure	General membrane	P_m
			Through thickness gradient	Q
	Junction with head or flange	Axial thermal gradient	Membrane	Q
			Bending	Q[2]
Any shell or head	Any section across entire vessel	External load or moment, or internal pressure	General membrane averaged across full section	P_m
		External load or moment	Bending across full section	P_m
	Near nozzle or other opening	External load or moment, or internal pressure	Local membrane	P_L
			Bending	Q
			Peak (fillet or corner)	F
	Any location	Temperature difference between shell and head	Membrane	Q
			Bending	Q
Dished head or conical head	Crown	Internal pressure	Membrane	P_m
			Bending	P_b
	Knuckle or junction to shell	Internal pressure	Membrane	P_L[3]
			Bending	Q
Flat head	Center region	Internal pressure	Membrane	P_m
			Bending	P_b
	Junction to shell	Internal pressure	Membrane	P_L
			Bending	Q[2]

Notes:
[1] Q and F classification of stresses refers to other than design condition.
[2] If the bending moment at the edge is required to maintain the bending stress in the middle to acceptable limits, the edge bending is classified as P_b. Otherwise it is classified as Q.
[3] Consideration should also be given to the possibility of wrinkling and excessive deformation in vessels with large diameter to thickness ratio.

strength. Therefore we have a factor of safety of 1.5 or 3 in terms of yield strength or ultimate strength, respectively. It is the purpose of the design codes that these multiples of either the yield or the ultimate strength are never exceeded in design.

The pressure vessel design codes often make specific recommendations on the limits depending on the conditions (or situations) of design. One typical such classification is in terms of design, normal, and upset (levels A and B), emergency (level C), faulted (level D) and test loadings, and accordingly limits are set appropriately. These stress limits have been discussed in some detail in Chapter 3. As an example, for design conditions the limits for the general primary membrane stress intensity, P_m, the local primary membrane stress intensity, P_L, and the combined membrane and bending stress intensity P_m (P_L) + P_b, are typically expressed as:

$$\begin{aligned} P_m &\leq S_m \\ P_L &\leq S_m \\ P_L + P_b &\leq 1.5S_m \end{aligned} \tag{4.3}$$

These limits are sometimes higher than the actual operating conditions. It is the intent of the design code that the limit on primary plus secondary stresses be applied to the actual operating conditions. For normal and upset conditions (sometimes indicated as levels A and B), the range of primary and secondary stresses, $P_L + P_b + Q$ is not allowed to exceed $3S_m$, or

$$|P_L + P_b + Q|_{range} \leq 3S_m \tag{4.4}$$

A word of caution is needed here. For example, a stress limit on some of the combination of stress categories such as $P_m(P_L) + P_b$, $P_L + Q$ needs to be carefully understood. The confusion arises because of the tendency to denote the stress intensity in a particular category by the symbol of that category, for example P is the stress intensity for the primary bending stress category. However ($P_L + P_b + Q$) is not the sum of the individual components of primary membrane, primary bending and secondary stress intensities. It is in fact the stress intensity evaluated from the principal stresses after the stresses from each category have been added together in the appropriate manner (that is not by adding the stress intensities). The primary plus secondary stress limits are intended to prevent excessive plastic deformation leading to incremental collapse, and to validate the application of elastic analysis when performing the fatigue evaluation. The limits ensure that the cycling of a load range results in elastic response of the material, also referred to as shakedown (when the ratcheting stops).

4.5 Special stress limits

The theories of failure have been outlined in Chapter 3, the important ones being the von Mises and the Tresca theories. However, none of the theories indicate any limit to the magnitude of the principal stresses, as long as their differences are within the specified limits (stress intensity limits). For uniform triaxial compressive stresses such a position is appropriate. However, for uniform triaxial tensile stresses, failures have been observed to occur, the predictions from the failure theories being that failures would not occur at all. From the available experimental data, it seems reasonable that limiting the mean of the principal stresses to the yield strength would ensure an adequate safety margin against failure, that is

$$\frac{\sigma_1 + \sigma_2 + \sigma_3}{3} < S_y \tag{4.5}$$

The ASME Boiler and Pressure Vessel Code takes a somewhat conservative estimate by limiting the mean of the principal stresses to 8/9 S_y, or

$$\frac{\sigma_1 + \sigma_2 + \sigma_3}{3} < \frac{8}{9} S_y \tag{4.6}$$

with an assumed value of the design stress intensity S_m equal to two-thirds S_y, or

$$S_m = \frac{2}{3} S_y \tag{4.7}$$

The limit on the sum of the principal stresses becomes equal to $4S_m$ or

$$\sigma_1 + \sigma_2 + \sigma_3 < 4S_m \tag{4.8}$$

4.6 Practical aspects of stress categorization

Hechmer and Hollinger have proposed ten guidelines on the evaluation of stresses in pressure vessels calculated by finite-element method in the spirit of the ASME Boiler and Pressure Vessel Code.[3]

1. Guideline 1 establishes the relationship between the membrane and bending stresses and the associated failure modes. The failure modes of concern are collapse/gross distortion (P_m), plastic collapse (P_b), excessive plastic deformation ($P_L + P_b$), and ratcheting and lack of shakedown ($P + Q$).
2. Guideline 2 relates to the first guideline by establishing the finite-element assessment for each failure mode. They maintain that for the

general primary membrane stress, P_m, finite-element analysis is not necessary, and simple equilibrium equations for net force and net moment are adequate. Finite-element analysis is an appropriate tool for evaluating the local primary membrane stress intensity, P_L, the primary membrane plus bending stress intensity, $P_L + P_b$, and the primary plus secondary stress intensity, $P + Q$. For bending of shell cross-sections (as shown in Figure 4.1), simple equations as in Eqs. (4.9) and (4.10) below are proposed

$$P_m = \frac{p}{A} \tag{4.9}$$

$$P_b = \frac{6m}{t^2} \tag{4.10}$$

Here p is the load and m the bending moment per unit length of the shell.

3. Guideline 3 defines stress classification lines (SCL) and stress classification planes (SCP) for the purpose of evaluating membrane and bending stresses. An SCL is shown in Figure 4.1.

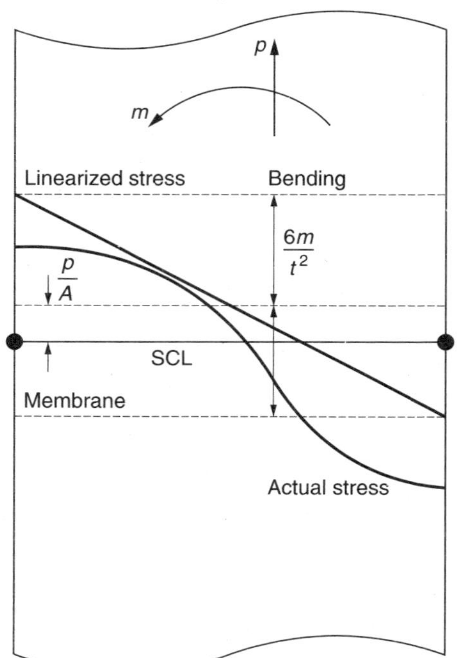

Figure 4.1 Stress classification line (SCL).

4. Guideline 4 establishes the global locations for assessment of stresses, and states that the general primary membrane stress intensity, P_m, should be evaluated remote from a discontinuity; whereas the primary membrane plus bending stress intensity, $P_L + P_b$, and primary plus secondary stress intensity, $P + Q$, should be evaluated at a discontinuity.
5. Guideline 5 establishes the criteria for local locations in terms of the SCL and SCP, and the orientation for the SCL–SCP caused by a discontinuity or blend radius. The blend radii are normally included in the finite-element model.
6. Guideline 6 provides the definition (rewording the code definition) of linearized stress as the stress represented by linear distributions which develop the same net forces and moments on a section as the total stress distribution.
7. Guideline 7 provides procedure for calculating membrane and bending stresses from component stresses, and not principal stresses. Linearized stresses are based on explicit computations by Eqs. (4.9) and (4.10) and not based on arbitrarily linearizing individual stress components.
8. Guideline 8 indicates the method for calculating principal stresses, stress intensities, and stress ranges. All six components of the stresses are linearized individually and then the principal stresses are calculated for the membrane stresses. For the bending stress only two of the component stresses are linearized, leaving the bending stress due to shear and through-thickness components.
9. Guideline 9 provides recommendation on the use of SCL and SCP for various geometries and states that for most axisymmetric situations, SCLs are appropriate. SCPs are recommended for special cases, such as flat plate with penetrations, and are deemed appropriate where the geometry has a well-defined plane that can be directly related to the failure mode.
10. Guideline 10 provides six fundamental recommendations on the evaluation of stresses by finite elements, which are:
    a. The FEA modeling techniques (mesh refinement, etc.) should be adequate for the level of accuracy needed for structural evaluation.
    b. The finite-element nodes should be such that the location of the SCLs can be readily established.
    c. An SCL or SCP may start or end at a singularity, because the integration of the loads along the line or the plane may override the effect of the singularity.
    d. Along discontinuities it is desirable to use an equilibrium type of analysis to obtain more accurate results.
    e. The failure locations and hence the locations of SCL and SCP should be established based on an overall review of the flow of stresses.

f. The methods recommended above generally apply for shells with R/t ratios greater than 4. For lower ratios of R/t, the level of accuracy is questionable.

4.7 Shape factor considerations

The primary stress intensity in the ASME Boiler and Pressure Vessel Code is intended to prevent uncontrolled plastic deformation and to provide a nominal factor of safety on the ductile burst pressure. These limits are based on the principles of limit design. The material is assumed to be elastic–perfectly plastic. For a straight bar in tension, a load producing yield stress, S_y, results in a collapse. If it is loaded in bending, collapse does not occur until the yield moment has been increased by the shape factor of the section.

The shape factor, α, is defined as the ratio of the load set producing a fully plastic section to the load producing initial yielding of the extreme fibers of the section. The shape factor for a rectangular section in pure bending is 1.5. The current stress intensity limits in the ASME Code rules are based on rectangular cross sections. For combined axial and bending loads, the load set to form a "plastic hinge" depends on the ratio of the tensile and bending loads.

An interaction curve as shown in Figure 4.1 is for a rectangular section. Note that when the membrane stress, P_m, is zero, the stress calculated elastically from the collapse moment for bending is $1.5S_y$. The factor 1.5 is the shape factor for the rectangular cross section. It should be noted that the current code limits are nonconservative for some sections. Such nonconservatism arises typically for sections with shape factors lower than 1.5 (such as an I-section with a thin web). Chattopadhyay has recommended design equations for different types of loadings to provide adequate safety for all combinations of axial and bending loads.[4] The proposed limits in Eqs. (4.11) to (4.13) are intended to replace the existing ones in the ASME Code, and apply to design, level C and testing limits as provided in Eqs (3.6) to (3.8) and shown pictorially in Figure 4.2. The proposed limits are shown below.

1. Design condition:

$$\begin{aligned} P_m &\le S_m \\ P_m + P_b &\le \alpha S_m \end{aligned} \tag{4.11}$$

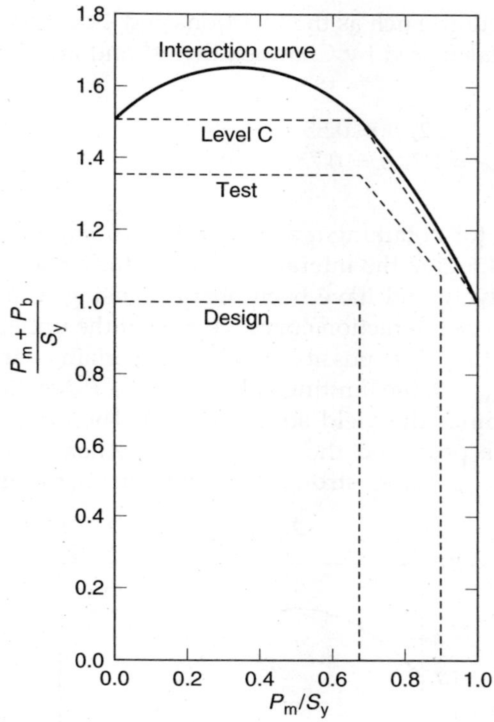

Figure 4.2 ASME Code limits.

2. Testing condition:

$$
\begin{aligned}
&P_m \le 0.9S_y \\
&P_m + P_b \le 0.9\alpha S_y, \text{ for} P_m \le 0.67S_y \\
&P_m + P_b \le [(0.9\alpha + 0.1\alpha\beta - \beta)/(1-\beta)]S_y - [(\alpha-1)/(1-\beta)]P_m
\end{aligned}
\tag{4.12}
$$

for $0.67S_y \le P_m \le 0.9S_y$

3. Level C condition:

$$
\begin{aligned}
&P_m = S_y \\
&P_L + P_b = \alpha S, \text{ for } P_L = 0.\beta S_y \\
&P_L + P_b = -[(\alpha-\beta)/(1-\beta)]P_L + [(\alpha-\beta)/(1-\beta)]S \\
&\qquad \text{for } \beta S_y = P_L = 0.S_y
\end{aligned}
\tag{4.13}
$$

where S is greater of $1.2S_m$ or S_y. The value of β is $1/\alpha$ for full sections identified as ones for which no abrupt changes in boundary occur.

For non full sections such as the I-sections and T-sections the values of α and β have been calculated by Chattopadhyay[4] and are:

- I Section: $\alpha = 1.2$, $\beta = 0.55$
- T Section: $\alpha = 1.7$, $\beta = 0.75$

The procedure for obtaining α and β will be discussed later.

Figures 4.3 to 4.6 give the interaction curves for circle, diamond, I- and T-sections respectively and have been obtained using techniques outlined in Appendix C. These interaction curves represent the upper limits of $(P_m + P_b)$ as a function of P_m. It is possible to observe certain characteristics in all these figures. At $P_m = 0$, the limiting value of $(P_m + P_b)$ is the shape factor of the cross-section times the yield strength (αS_y). Then $(P_m + P_b)$ increases with P_m, reaches a peak and then drops to S_y when P_m reaches S_y. The peak value of $(P_m + P_b)$ is strongly dependent on the geometry of the

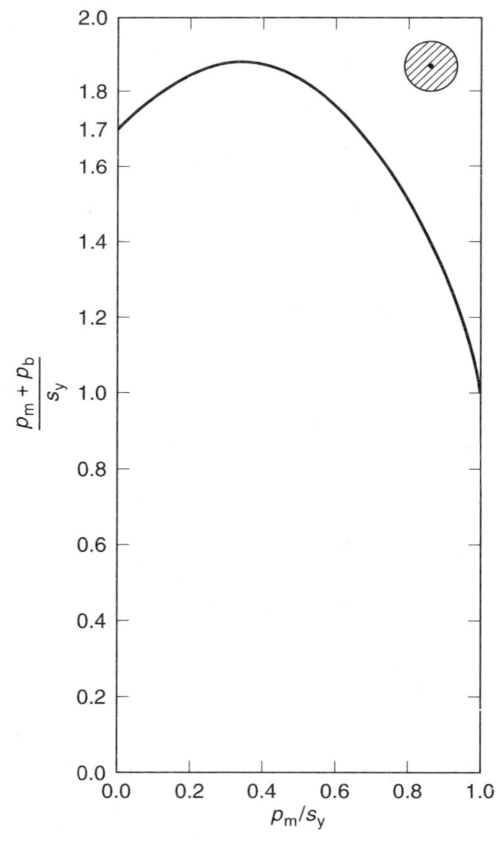

Figure 4.3 Interaction curve for a circular cross-section.

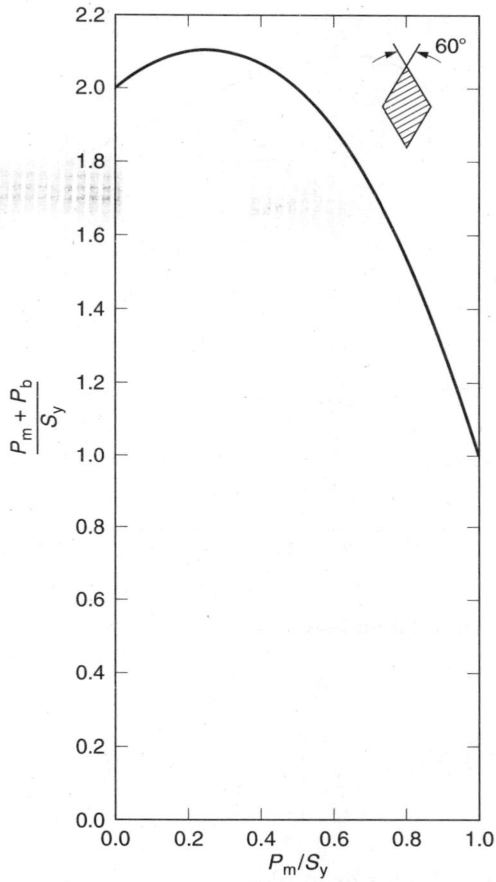

Figure 4.4 Interaction curve for a diamond cross-section.

cross-section. Another feature that is noteworthy is that different cross-sections can have identical shape factors, but differ significantly in the interaction characteristics (see Figures 4.3 and 4.6).

For rectangular cross-sections, a closed-form mathematical expression describes the interaction curve. For an arbitrary cross-section, no such expression exists, and a large number of computations are necessary to obtain the interaction curve at discrete values of P_m and $(P_m + P_b)$. For certain sections where no abrupt changes in boundary occur, the interaction curves may be approximated by analytical expressions involving the shape factor, α. Such cross sections may be referred to as "full" sections. (Note that an I- or a T-section cannot be called a "full" section). Examples of full sections include rectangular, circular, diamond and trapezoidal sections.

The interaction curve for a rectangular section as shown in Figure 4.2 can be mathematically described as

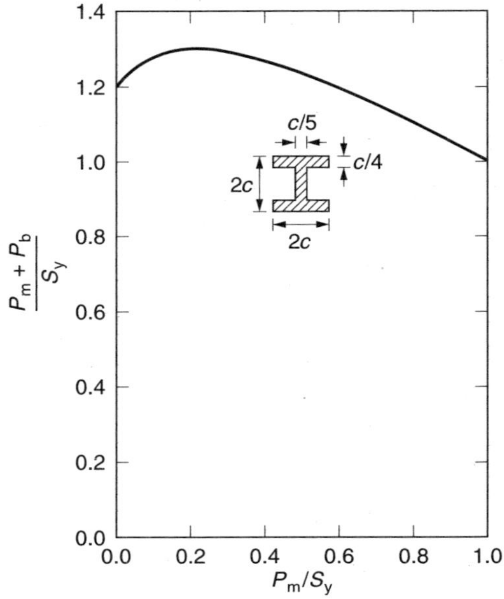

Figure 4.5 Interaction curve for an I-section.

$$\frac{P_{\mathrm{m}} + P_{\mathrm{b}}}{S_{\mathrm{y}}} = 1.5\left[1 - \left(\frac{P_{\mathrm{m}}}{S_{\mathrm{y}}}\right)^2\right] + \frac{P_{\mathrm{m}}}{S_{\mathrm{y}}} \tag{4.14}$$

See References 5 and 6 for examples.

For nonrectangular sections, we can tentatively replace 1.5 and check the validity of this substitution later. We therefore have

$$\frac{P_{\mathrm{m}} + P_{\mathrm{b}}}{S_{\mathrm{y}}} = \alpha\left[1 - \left(\frac{P_{\mathrm{m}}}{S_{\mathrm{y}}}\right)^2\right] + \frac{P_{\mathrm{m}}}{S_{\mathrm{y}}} \tag{4.15}$$

Using the notation

$$y = \frac{P_{\mathrm{m}} + P_{\mathrm{b}}}{S_{\mathrm{y}}}$$

and

$$x = \frac{P_{\mathrm{m}}}{S_{\mathrm{y}}}$$

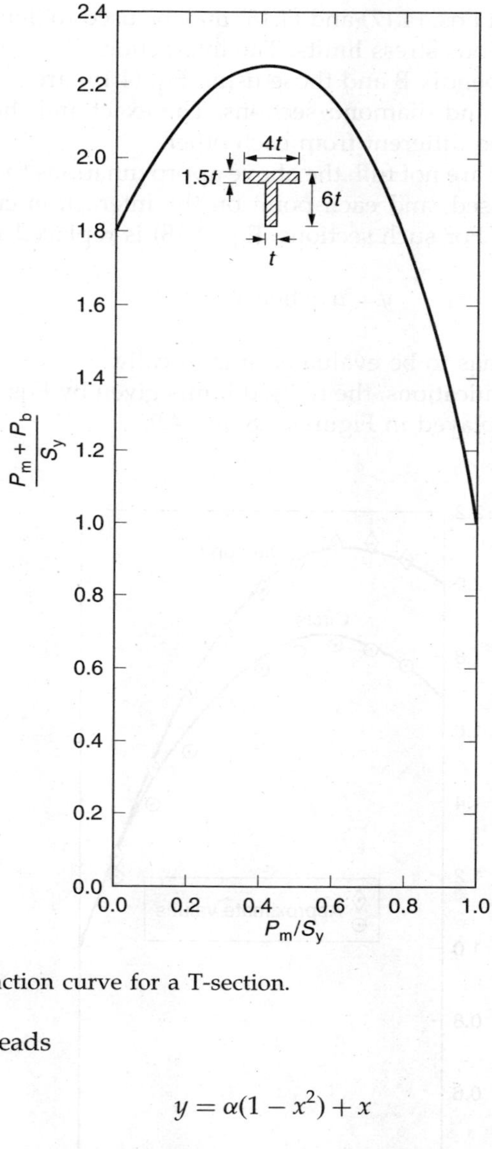

Figure 4.6 Interaction curve for a T-section.

Eq. (4.15) reads

$$y = \alpha(1 - x^2) + x \tag{4.16}$$

y is a maximum when $dy/dx = 0$, which gives

$$y_{\max} = \alpha + \frac{1}{4\alpha} \text{ at } x = \frac{1}{2\alpha} \tag{4.17}$$

Furthermore,

$$y = \alpha \text{ when } x = 0, \frac{1}{\alpha} \tag{4.18}$$

The results from Eqs. (4.17) and (4.18) may be used to define domains of applicability of various stress limits. The interaction curves based on exact methods using Appendix B and those using Eq. (4.15) are shown in Figure 4.7 for the circular and diamond sections. The exact and the approximate solutions are not too different from each other.

For sections that are not full, the above approximations to the interaction curves cannot be used, and each point on the interaction curve has to be obtained explicitly. For such sections, Eq. (4.18) is replaced with

$$y = \alpha \text{ when } x = 0, \beta \tag{4.19}$$

The value of β has to be evaluated numerically.

With these modifications, the revised limits given by Eqs. (4.11) to (4.13) are graphically displayed in Figures 4.8 and 4.9.

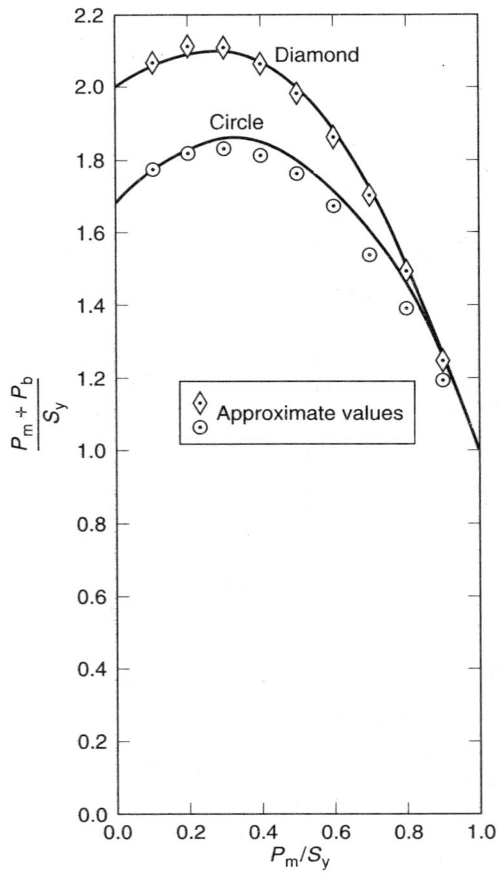

Figure 4.7 Comparison of exact and approximate interaction curves.

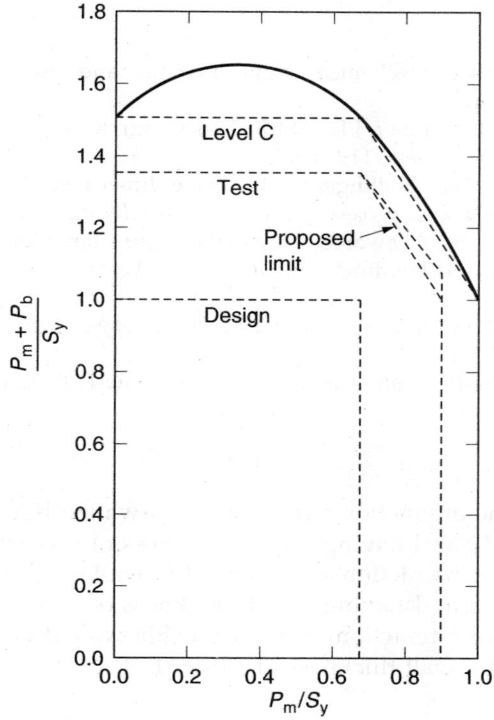

Figure 4.8 Proposed stress limits for rectangular cross-sections.

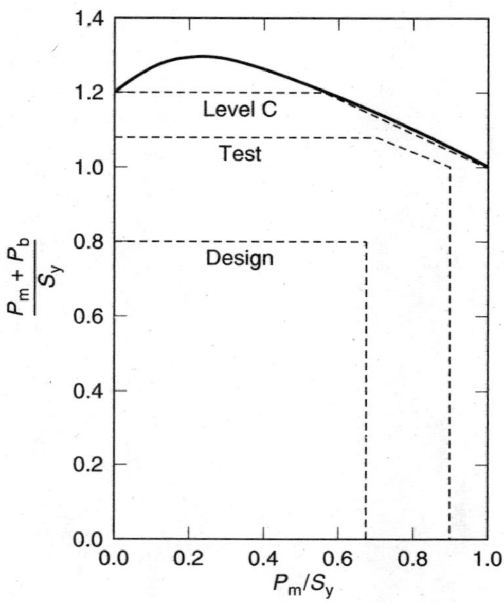

Figure 4.9 Proposed stress limits for an I-section.

References

1. American Society of Mechanical Engineers, Boiler and Pressure Vessel Code, ASME, New York.
2. Pastor, T.P., and Hechmer, J.L., ASME Task Group Report on Primary Stress, *J. Pressure Vessel Technol.*, 119, 1997.
3. Hechmer, J.L., and Hollinger, G.L. Three-dimensional stress criteria – summary of the PVRC project, *J. Pressure Vessel Technol.*, 122, 105–109, 2000.
4. Chattopadhyay, S., Allowable stresses for nonrectangular sections under combined axial and bending loads, *J. Pressure Vessel Technol.*, 110, 188–193, 1988.
5. Seely, F.B., and Smith, J.O., *Advanced Mechanics of Materials*, 2nd ed., Wiley, New York, 1952.
6. Hodge, P.G., *Plastic Analysis of Structures*, McGraw-Hill, New York, 1959.

Problems

1. Determine the interaction curve for a thin-walled I-beam of height h and width $h/2$ and having a wall thickness of t ($t<<h$).
2. Determine the interaction curve for a thin-walled T-beam of height h and width $h/2$ and having a wall thickness of t ($t<<h$).
3. Determine the interaction curve for a thin-walled circular beam of radius R and a wall thickness of t ($t<<h$).

chapter five

Design of cylindrical shells

Contents

5.1 Introduction

Cylindrical shells are used in nuclear, fossil and petrochemical industries. They are also used in heat exchangers of the shell and tube type. Generally these vessels are easy to fabricate and install and economical to maintain. The design procedures in pressure vessel codes for cylindrical shells are mostly based on linear elastic assumption, occasionally allowing for limited inelastic behavior over a localized region. The shell thickness is the major design parameter and is usually controlled by internal pressure and sometimes by external pressure which can produce buckling. Applied loads are also important in controlling thickness and so are the discontinuity and thermal stresses. The basic thicknesses of cylindrical shells are based on simplified stress analysis and allowable stress for the material of construction. There are some variations of the basic equations in various design codes. Some of the equations are based on thick-wall Lame equations. In this chapter such equations will be discussed. Also we shall discuss the case of cylindrical shells under external pressure where there is a propensity of buckling or collapse.

5.2 Thin-shell equations

A shell is a curved plate-type structure. We shall limit our discussion to shells of revolutions. Referring to Figure 5.1 this is denoted by an angle φ, the meridional radius r_1 and the conical radius r_2, from the center line. The horizontal radius when the axis is vertical is r.

If the shell thickness is t, with z being the coordinate across the thickness, following the convention of Flugge,[1] we have the following stress resultants:

$$N_\theta = \int_{-\frac{t}{2}}^{\frac{t}{2}} \sigma_\theta \left(\frac{r_1 + z}{r_1}\right) dz \tag{5.1}$$

$$N_\phi = \int_{-\frac{t}{2}}^{\frac{t}{2}} \sigma_\phi \left(\frac{r_2 + z}{r_2}\right) dz \tag{5.2}$$

$$N_{\theta\phi} = \int_{-\frac{t}{2}}^{\frac{t}{2}} \sigma_{\theta\phi} \left(\frac{r_2 + z}{r_2}\right) dz \tag{5.3}$$

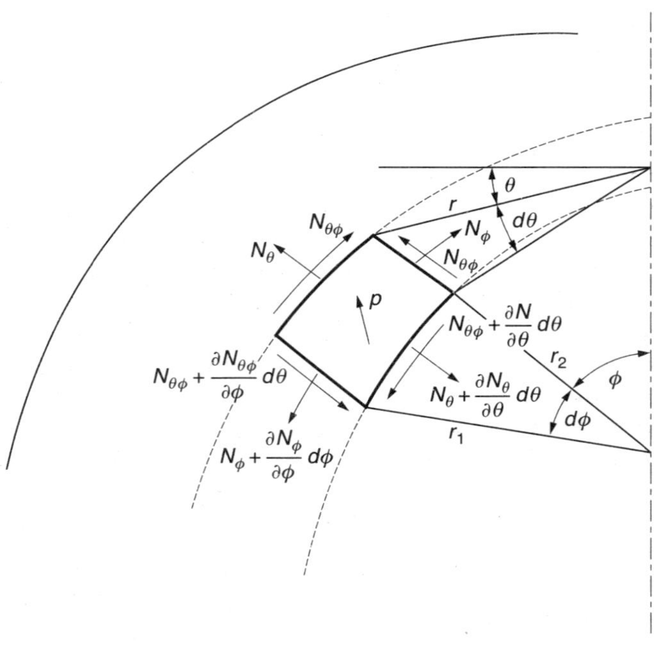

Figure 5.1 Thin shell of revolution.

$$N_{\phi\theta} = \int_{-\frac{t}{2}}^{\frac{t}{2}} \sigma_{\phi\theta}\left(\frac{r_1 + z}{r_1}\right)dz \tag{5.4}$$

These stress resultants are assumed to be due only to an internal pressure, p, acting in the direction of r. For membrane shells where the effects of bending can be ignored, all the moments are zero and further development leads to

$$N_{\phi\theta} = N_{\theta\phi} \tag{5.5}$$

The following equations result from considering force equilibrium along with the additional requirement of rotational symmetry:

$$\frac{d(rN_\phi)}{d\phi} - r_1 N_\theta \cos\phi = 0 \tag{5.6}$$

$$N_\theta = pr_2 - \frac{r_2}{r_1} N_\phi \tag{5.7}$$

Noting that $r = r_2 \sin\phi$, we have, by solving Eqs. (5.6) and (5.7),

$$N_\phi = \frac{pr_2}{2} \tag{5.8}$$

$$N_\theta = \frac{pr_2}{2}(2 - \frac{r_2}{r_1}) \tag{5.9}$$

The above two equations are the results for a general shell of revolution. Two specific cases result:

1. For a spherical shell of radius R, $r_1 = r_2 = R$, which gives

$$N_\phi = N_\theta = \frac{pR}{2} \tag{5.10}$$

2. For a cylindrical pressure vessel of radius R, we have $r_1 = \infty$, $r_2 = R$, which gives

$$N_\phi = \frac{pR}{2} \tag{5.11}$$

$$N_\theta = pR \tag{5.12}$$

This gives the hoop stress

$$\sigma_{\text{hoop}} = \sigma_\theta = \frac{N_\theta}{t} = \frac{pR}{t} \tag{5.13}$$

and the longitudinal stress

$$\sigma_{\text{long}} = \sigma_\phi = \frac{N_\phi}{t} = \frac{pR}{2t} \tag{5.14}$$

These results will be shown to be identical to the results that follow.

Let us consider a long thin cylindrical shell of radius R and thickness t, subject to an internal pressure p. By thin shells we mean the ones having the ratio R/t typically greater than about 10. If the ends of the cylindrical shell are closed, there will be stresses in the hoop as well as the axial (longitudinal) directions.

A section of such a shell is shown in Figure 5.2. The hoop (circumferential) stress, σ_{hoop} and the longitudinal stress, σ_{long} are indicated in the figure. The shell is assumed to be long and thin resulting in σ_{hoop} and σ_{long} to be uniform through the thickness. Therefore in this case σ_{hoop} and σ_{long} are also referred to as membrane stress (there are no bending stresses associated with this type of loading).

Considering equilibrium across the cut section, we have,

$$pL\,(2R) = 2\sigma_{\text{hoop}}\,tL$$

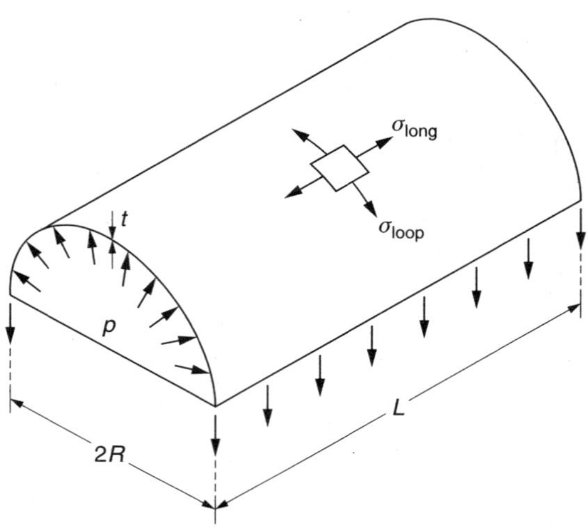

Figure 5.2 Thin cylindrical shell.

which gives

$$\sigma_{\text{hoop}} = \frac{pR}{t} \tag{5.15}$$

Considering a cross-section of the shell perpendicular to its axis, we have

$$p\,\pi R^2 = \sigma_{\text{long}}\,(2\pi R t)$$

which gives

$$\sigma_{\text{hoop}} = \frac{pR}{2t} \tag{5.16}$$

5.3 Thick-shell equations

For R/t ratios typically less than 10, Eqs. (5.15) and (5.16) tend not to be accurate, and thick-shell equations have to be used.

Consider a thick cylindrical shell of inside radius R_i and outside radius R_o subjected to an internal pressure p as shown in Figure 5.3.

The stress function for this case (refer to Appendix I) is given as a function of radius r as

$$\Phi = A \ln r + Br^2 \tag{5.17}$$

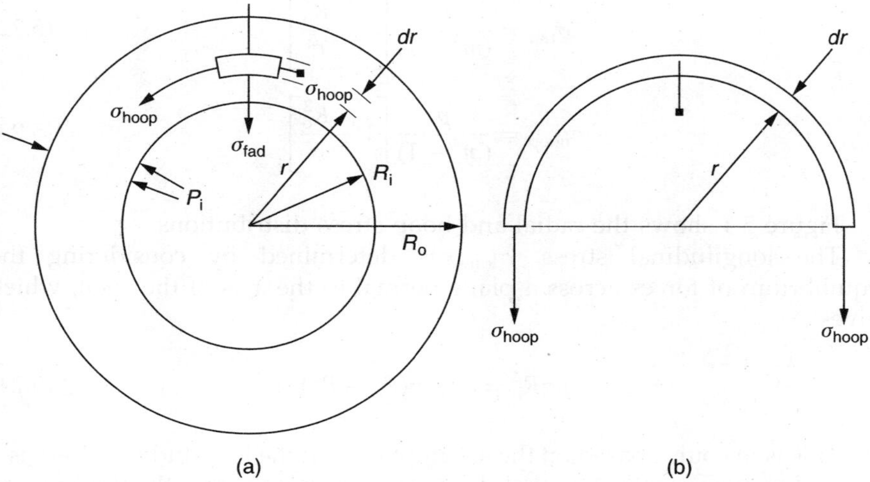

Figure 5.3 Thick cylindrical shell.

with A and B to be determined by the boundary conditions.

If we indicate the radial stress as σ_{rad} and the hoop and longitudinal stress as indicated previously by σ_{hoop} and σ_{long}, we have

$$o_{\text{rad}} = \frac{1}{r}\frac{d\Phi}{dr} = \frac{A}{r^2} + 2B \tag{5.18}$$

$$o_{\text{hoop}} = \frac{d^2\Phi}{dr^2} = -\frac{A}{r^2} + 2B \tag{5.19}$$

The constants A and B are determined from the following boundary conditions:

$$\sigma_{\text{rad}} = -p \quad \text{at} \quad r = R_i,$$

$$\sigma_{\text{rad}} = 0 \quad \text{at} \quad r = R_{\text{o}} \tag{5.20}$$

Substituting (5.20) into (5.18) and (5.19), we have

$$A = -\frac{R_{\text{i}}^2 R_{\text{o}}^2 p}{(R_{\text{o}}^2 - R_{\text{i}}^2)}$$

$$B = \frac{R_{\text{i}}^2 p}{2(R_{\text{o}}^2 - R_{\text{i}}^2)} \tag{5.21}$$

Denoting the ratio of the outside to inside radii as m, so that $m = R_{\text{o}}/R_{\text{i}}$, we obtain the radial and hoop stresses

$$\sigma_{\text{rad}} = \frac{p}{(m^2 - 1)}\left[1 - \frac{R_{\text{o}}^2}{r^2}\right] \tag{5.22}$$

$$\sigma_{\text{hoop}} = \frac{p}{(m^2 - 1)}\left[1 + \frac{R_{\text{o}}^2}{r^2}\right] \tag{5.23}$$

Figure 5.4 shows the radial and hoop stress distributions.

The longitudinal stress, σ_{long} is determined by considering the equilibrium of forces across a plane normal to the axis of the shell, which gives

$$p\pi R_{\text{i}}^2 = \sigma_{\text{long}}\pi(R_{\text{o}}^2 - R_{\text{i}}^2) \tag{5.24}$$

This is of course based on the assumption that the longitudinal stress is a form of membrane stress in that there is no variation across the thickness of the shell. Thus we have

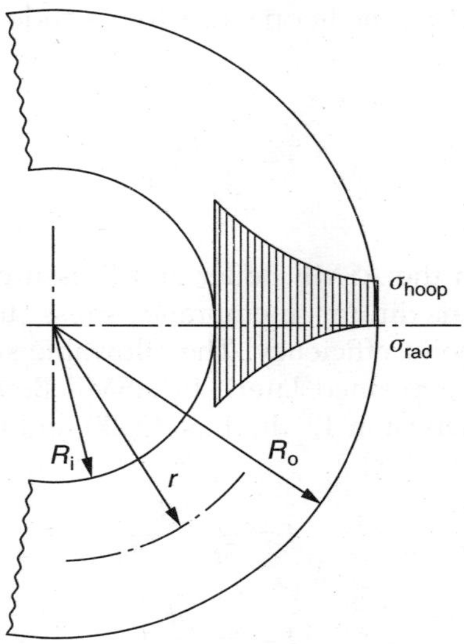

Figure 5.4 Hoop and radial stress distribution.

$$\sigma_{long} = \frac{pR_i^2}{(R_o^2 - R_i^2)} = \frac{p}{(m^2 - 1)} \tag{5.25}$$

It should be noted however that the solutions indicated by Eqs. (5.22), (5.23), and (5.25) are valid for regions remote from discontinuities.

5.4 *Approximate equations*

For a moderately thick shell employing thin-shell theory and using the mean radius R_m we get the expression of the hoop stress, σ_{hoop}, as

$$\sigma_{hoop} = \frac{pR_m}{t} = \frac{p(R_i + t/2)}{t} \tag{5.26}$$

Equating the hoop stress, σ_{hoop}, to the code-allowable design stress, S_m, we have

$$t = \frac{pR_i}{S_m - 0.5p} \tag{5.27}$$

Rewriting Eq. (5.27) in terms of the outside radius, R_o we have,

$$\sigma_{hoop} = \frac{pR_m}{t} = \frac{p(R_o - t/2)}{t} \tag{5.28}$$

Once again equating the hoop stress to the code-allowable design stress, S, we have

$$t = \frac{pR_o}{S_m + 0.5p} \tag{5.29}$$

The equations in the ASME Boiler and Pressure Vessel Code are based on equating the maximum membrane stress to the allowable stress corrected for weld joint efficiency. The allowable stress, S_m, is replaced by the term SE (to be explained later). In ASME Boiler and Pressure Vessel Code, Section VIII Division 1,[2] the Eqs. (5.27) and (5.29) are modified as:

$$t = \frac{pR}{SE - 0.5p} \tag{5.30}$$

$$t = \frac{pR}{SE + 0.5p} \tag{5.31}$$

In ASME Boiler and Pressure Vessel Code Section III, Division 1,[2] the equations used are

$$t = \frac{pR}{SE - 0.6p} \tag{5.32}$$

$$t = \frac{pR}{SE + 0.4p} \tag{5.33}$$

In Eqs. (5.30) and (5.32), R stands for the inside radius, R_i, whereas in Eqs. (5.31) and (5.33) it stands for the outside radius, R_o. In both of the above equations, S is the allowable stress and E is the joint efficiency. This joint efficiency is employed because cylindrical shells are often fabricated by welding. The values of E depend on the type of radiographic examination performed at various welded seams of the shell.

5.5 Buckling of cylindrical shells

Consider a long, thin cylindrical shell of mean diameter D and wall thickness t subjected to an external pressure P. The cylinder is in a stable configuration as long as it remains circular in shape. If there is an initial ellipticity, the cylinder will be in an unstable condition and will eventually buckle.[4]

If the cylinder is sufficiently long, the end effects may be neglected and the problem may be considered as two-dimensional.

Summing up the forces in the radial direction, we have

$$\left(N + \frac{dN}{dS}ds\right)\frac{d\theta}{2} + N\frac{d\theta}{2} + \left(V + \frac{dv}{ds}ds\right) - V - Pds = 0$$

or

$$Nd\theta + \frac{dv}{ds}ds - Pds = 0$$

or

$$Nd\theta + \frac{dv}{ds}Rd\theta - PRd\theta = 0, (R = D/2)$$

$$P - \frac{dV}{dS} - \frac{N}{R} = 0$$

Summation of forces in the tangential direction (see Figure 5.5) gives

$$-\left(N + \frac{dN}{ds}ds\right) + N + \left(v + \frac{dv}{ds}\right)\frac{d\theta}{2} + v\frac{d\theta}{2} = 0$$

$$-\frac{dN}{ds}ds + vd\theta = 0$$

$$V - R\frac{dN}{ds} = 0$$

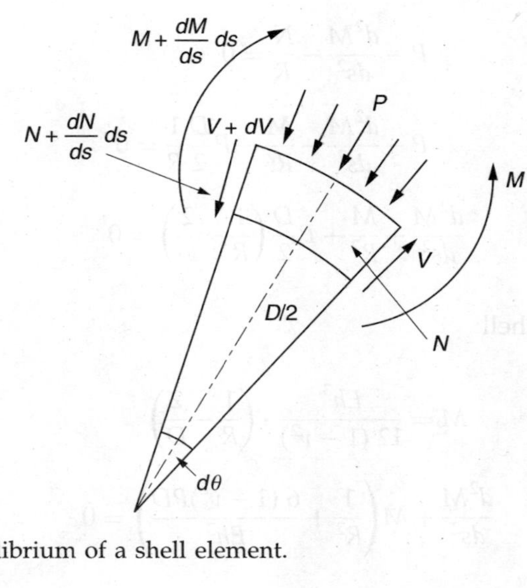

***Figure* 5.5** Equilibrium of a shell element.

With $V = \frac{dM}{ds}$, we have

$$\frac{dN}{ds} = \frac{1}{R}\frac{dM}{ds}$$

Assuming deviation from circular shape to be small, we have

$$N = \frac{M}{R} + C$$

where C is a constant.

When

$$R = \frac{D}{2},$$

$$M = V = 0$$

and

$$N = P\frac{D}{2}$$

Therefore

$$N = \frac{M}{R} + P\frac{D}{2}$$

$$P - \frac{d^2M}{ds^2} - \frac{N}{R} = 0$$

$$P - \frac{d^2M}{ds^2} - \frac{M}{R^2} - P\frac{D}{2}\frac{1}{R} = 0$$

$$\frac{d^2M}{ds^2} + \frac{M}{R^2} + P\frac{D}{2}\left(\frac{1}{R} - \frac{2}{D}\right) = 0$$

For a curved shell:

$$M = \frac{Eh^3}{12\,(1 - \nu^2)} \cdot \left(\frac{1}{R} - \frac{2}{D}\right)$$

$$\frac{d^2M}{ds^2} + M\left(\frac{1}{R^2} + \frac{6\,(1 - \nu^2)PD}{Eh^3}\right) = 0$$

with $R \cong \frac{D}{2}$

$$\frac{d^2M}{ds^2} + \left(\frac{4}{D^2} + \frac{6PD\,(1-\nu^2)}{Eh^3}\right) M = 0$$

with $k^2 = \frac{4}{D^2} + \frac{6PD(1-\nu^2)}{Eh^3}$

$$\frac{d^2M}{ds^2} + k^2M = 0$$

$$M = C_1 \sin ks + C_2 \cos ks$$

$$\frac{dM}{ds} = C_1 k \cos ks + C_2 k \sin ks$$

$$\frac{dM}{ds} = 0 \text{ at s} = 0$$

$$\frac{dM}{ds} = 0 \text{ at s} = \frac{\pi D}{4}$$

$$C_1 = 0$$

$$\sin\left(\frac{k\pi D}{4}\right) = 0$$

$$\left(\frac{k\pi D}{4}\right) = n\pi$$

or

$$kD = 4n$$

$$\frac{16n^2}{D^2} = \frac{4}{D^2} + \frac{6P_{cr}D(1-\nu^2)}{Eh^3}$$

or

$$\frac{6P_{cr}D(1-\nu^2)}{Eh^3} = \frac{4}{D^2}(4n^2-1)$$

With a minimum value of $n = 1$, we get

$$P_{cr} = \frac{2E}{(1-\nu^2)}\left(\frac{t}{d}\right)^3 \tag{5.34}$$

For cylinder with shorter lengths, where the ends are free to expand axially and rotate with the restriction of expanding radially, the critical pressure is given by[3]

$$P_{0_{cr}} = \frac{2E}{3(1-\nu^2)}\left(\frac{t}{D}\right)^3 \left[(n^2-1) + \frac{2n^2-1-\nu}{1+\frac{4n^2L^2}{\pi^2 D^2}}\right] + \left[2E\left(\frac{t}{D}\right)\frac{1}{(n^2-1)(1+\frac{4n^2L^2}{\pi^2 D^2}}\right]$$

When $\frac{l}{D}$ is large:

$$P_{0_{cr}} = \frac{2}{3}\frac{E}{(1-\nu^2)}\left(\frac{t}{D}\right)^3 (n^2-1) \tag{5.35}$$

so that it becomes identical to the buckling pressure in Eq. (5.34) for $n = 2$.

In the ASME code, the critical pressure is calculated for two situations, involving the ratio of the outside diameter to the thickness (D_o/t)

1. $\frac{D_o}{t} \geq 10$
2. $\frac{D_o}{t} < 10$

The basic Eq. (5.19) is modified to include inelastic buckling. For the first case above, a factor of safety of 3.0 is used. For the second case, a variable factor of safety is used starting with a factor of safety of 3.0 for $D_o/t = 10$ to 2.0 for $D_o/t = 4$. As the cylinder becomes progressively thicker, the buckling ceases to be a plausible mode of failure. The ASME procedure is an involved one in which two sets of curves have to be used to investigate buckling. The procedure becomes complicated for large D_o/t, where checks need to be made whether the buckling is in the elastic region or in the plastic region.

5.6 Discontinuity stresses in pressure vessels

Let us take the special case of discontinuity at a juncture between a cylindrical vessel and a hemispherical head subjected to internal pressure p. For simplicity let us assume the spherical head and the cylindrical shell are of the same thickness. If the mean radius and the thickness of the shell are denoted by R_m and t respectively, then the hoop and the longitudinal stresses in the cylindrical shell are given by:

$$\sigma^{c}_{hoop} = \frac{pR_m}{t} \tag{5.36}$$

$$\sigma^{c}_{long} = \frac{pR_m}{2t} \tag{5.37}$$

The hoop and the longitudinal stresses in the spherical shell are given by:

$$\sigma^{s}_{hoop} = \frac{pR_m}{2t} \tag{5.38}$$

$$\sigma^{s}_{long} = \frac{pR_m}{2t} \tag{5.39}$$

The radial growth or dilation of the cylindrical shell under internal pressure p is given by

$$\delta^{c}_{r} = \frac{pR_m^2}{2Et}(2 - \nu) \tag{5.40}$$

That of the spherical region is given by

$$\delta^{s}_{r} = \frac{pR_m^2}{2Et}(1 - \nu) \tag{5.41}$$

where ν is the Poisson's ratio.

If the spherical and the cylindrical portions were separated, the difference in the radial growth would be

$$\delta_r = \delta^{c}_{r} - \delta^{s}_{r} = \frac{pR_m^2}{2Et} \tag{5.42}$$

In the actual vessel the hemispherical head and the cylindrical shell are kept in place by shear force, V and moment M per unit circumference. These discontinuity forces produce local bending stresses in the adjacent portions of the vessel. The deflection and the slope induced at the edges of the cylindrical and spherical portions by the force V are equal. The continuity at the juncture will be satisfied if M equals zero and V is such that it produces a deflection of $\delta/2$.

Applying the results from semi-infinite beam on an elastic foundation due to M and V,[5] and substituting the spring rate of the foundation by (Et/R_m^2):

$$\delta = \frac{2V\beta R_m^2}{Et} e^{-\beta x}\cos(\beta x) - \frac{2M\beta^2 R_m^2}{Et} e^{-\beta x}(\cos(\beta x) - \sin(\beta x)) \tag{5.43}$$

where β is the attenuation factor, given by

$$\beta = \sqrt[4]{\frac{3(1-\nu^2)}{R_m^2 t^2}} \tag{5.44}$$

We have with $\delta = \delta/2$ and $M = 0$ at $x = 0$,

$$\frac{\delta}{2} = \frac{2V\beta R_m^2}{Et} \tag{5.45}$$

Substituting the value of δ from Eq. (5.45) we have

$$V = \frac{p}{8\beta} \tag{5.46}$$

The longitudinal stress and the hoop stress distribution in the cylindrical region is then given by

$$\sigma_{long}^{c} = \frac{pR_m}{2t} \pm \frac{6}{t^2}\frac{p}{8\beta^2}e^{-\beta x}\sin(\beta x) \tag{5.47}$$

$$\sigma_{hoop}^{c} = \frac{pR_m}{t} - \frac{pR_m}{4t}e^{-\beta x}\cos(\beta x) \pm \frac{3\nu p}{4t^2\beta^2}e^{-\beta x}\sin(\beta x) \tag{5.48}$$

Using numerical values as, $p = 2$ MPa, $R_m = 1$ m, $t = 25$ mm, and Poisson's ratio, $\nu = 0.3$, we have from Eq. (5.44), $\beta = 0.008127/\text{mm}$ and the longitudinal and the hoop stresses become

$$\sigma_{long}^{c} = 40 \pm 36e^{-\beta x}\sin(\beta x) \quad \text{MPa} \tag{5.49}$$

$$\sigma_{hoop}^{c} = 80 - 20e^{-\beta x}\cos(\beta x) \pm 10.9e^{-\beta x}\sin(\beta x) \quad \text{MPa} \tag{5.50}$$

In Eq. (5.49) the first quantity – the membrane longitudinal stress – is a constant (equal to 40 MPa) along the length of the vessel, while the second quantity – the bending stress – varies along the length. In Eq. (5.50) the membrane hoop stress (equal to 80 MPa) stays constant along the length, while the direct compression stress due to shortening of the radius and the bending stress varies along the length of the vessel.

References

1. Flugge, W., *Stresses in Shells*, Springer-Verlag, New York, 1960.
2. American Society of Mechanical Engineers, Boiler and Pressure Vessel Code, ASME, New York.
3. Windenberg, D.P. and Trilling, C., Pressure vessel and piping design, *Collected Papers*, American Society of Mechanical Engineers, New York, 1960.

4. Timoshenko, S.P., and Gere, J.M., *Theory of Elastic Stability*, 2nd ed., McGraw-Hill, New York, 1961.
5. Hetenyi, M., *Beams on Elastic Foundation*, University of Michigan Press, Ann Arbor, 1958.

Problems

1. Find the thickness of a cylindrical shell 2 m in diameter if it is required to contain an internal pressure of 7 MPa. The allowable stress in the material is 140 MPa.
2. A thick cylindrical shell of 1.2 m inside diameter and 1.5 m outside diameter is subjected to an internal pressure of 35 MPa. Determine the following:
   a. Magnitude and location of the maximum hoop stress
   b. Magnitude of the maximum radial stress and its location
   c. Average hoop stress
3. A thick cylinder has an inside diameter of 300 mm and an outside diameter of 450 mm. If the allowable stress is 175 MPa, what is the maximum internal pressure that can be applied?
4. A cylinder has an inside radius of 1.8 m and is subjected to an internal pressure of 0.35 MPa. What is the required thickness if the allowable stress is 105 MPa?
5. For problem 4, what is the required thickness if thick cylinder equations were used?
6. Using ASME Boiler and Pressure Vessel Code equations, determine the thickness of a cast-iron pressure vessel subjected to an internal pressure of 0.5 MPa using a joint efficiency of 85 percent and a corrosion allowance of 1.5 mm. The allowable stress (S_m) of the material is 14 MPa.
7. The inside diameter of a boiler made of alloy steel is 2 m. The internal pressure is 0.75 MPa. The allowable stress is 140 MPa and the joint efficiency is 70 percent. What thickness is to be used? Use ASME Code equations.
8. A thick vessel is to be designed to withstand an internal pressure of 50 MPa. An internal diameter of 300 mm is specified and steel with a yield stress of 183 MPa is to be used. Calculate the wall thickness using the Tresca and von Mises criteria using a factor of safety of 1.5.

chapter six

Design of heads and covers

Contents

6.1 Introduction

Heads are one of the important parts in pressure vessels and refer to the parts of the vessel that confine the shell from below, above, and the sides. The ends of the vessels are closed by means of heads before putting them into operation.

The heads are normally made from the same material as the shell and may be welded to the shell itself. They also may be integral with the shell in forged or cast construction. The head geometrical design is dependent on the geometry of the shell as well as other design parameters such as operating temperature and pressure.

The heads may be of various types such as:

- Flanged
- Ellipsoidal
- Torispherical
- Hemispherical
- Conical
- Toriconical

The different types of heads are shown in Figure 6.1.

The geometry of the head is selected based on the function as well as on economic considerations, and methods of forming and space requirements. The elliptical and torispherical heads are most commonly used. The carbon steel hemispherical heads are not so economical because of the high manufacturing costs associated with them. They are thinner than the cylindrical shell to which they are attached, and require a smooth transition between the two to avoid stress concentration effects.

The thickness values of the elliptical and torispherical heads are typically the same as the cylindrical shell sections to which they are attached. Conical and toriconical heads are used in hoppers and towers.

6.2 Hemispherical heads under internal pressure

The force due to internal pressure is resisted by the membrane stress in the shell (see Figure 6.2). Because of the geometrical symmetry, the membrane stresses in the circumferential and the meridional directions are the same, and are denoted by S. We have

$$P\pi R^2 = 2\pi RSt$$

and

$$S = \frac{PR}{2t} \tag{6.1}$$

where S is the membrane stress, $(S_\theta = S_\phi)$, from symmetry.

The hoop and meridional strains are indicated by ε_θ and ε_ϕ

$$\varepsilon_\phi = \frac{w}{R} = \varepsilon_\theta \tag{6.2}$$

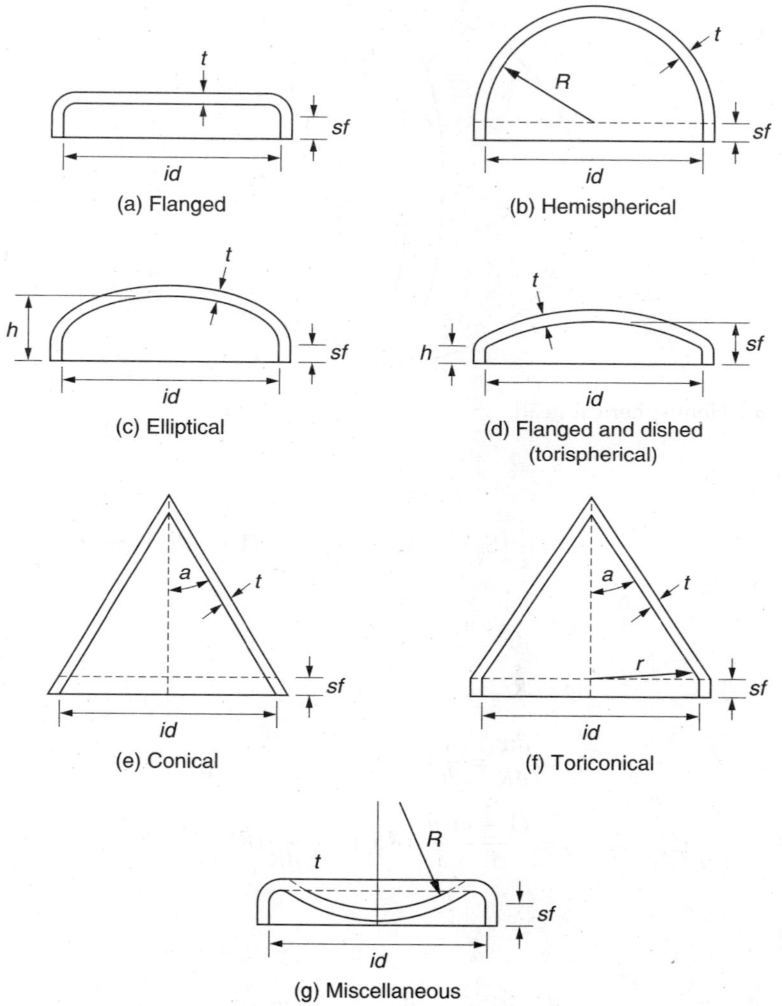

Figure 6.1 Different types of heads.
(Modified from ASME Boiler and Pressure Vessel Code, ASME, New York.)

where w is the radial displacement, and

$$\varepsilon_R = \frac{dw}{dR} \tag{6.3}$$

The stress–strain relationship is given by

$$\varepsilon_R = \frac{1}{E}\left[S_R - \nu(S_\theta + S_\phi)\right] = \frac{1}{E}[S_R - 2\nu S_\theta] \tag{6.4}$$

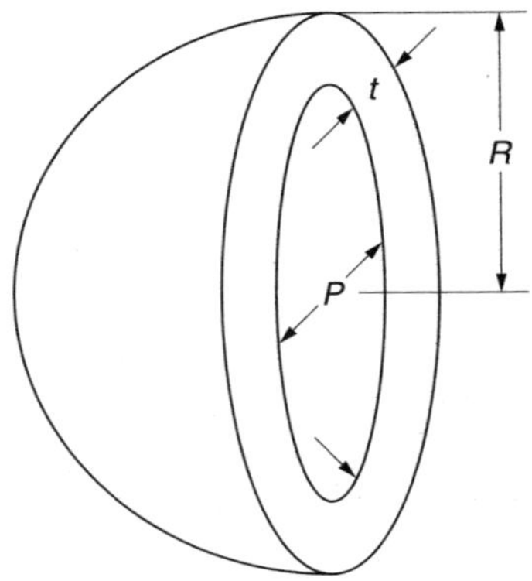

Figure 6.2 Hemispherical head.

$$\varepsilon_\theta = \frac{1}{E}\left[S_\theta - \nu(S_R + S_\phi\right] = \frac{1}{E}[(1-\nu)S_\theta - \nu S_R] \tag{6.5}$$

$$\varepsilon_R = \frac{dw}{dR};\ \varepsilon_\theta = \frac{w}{R}$$

$$w = R\varepsilon_\theta$$

$$\varepsilon_R = \frac{dw}{dR} = \frac{d}{dr}(R\varepsilon_\theta)$$

$$\frac{1}{\varepsilon}[S_R - 2\nu S_\theta] = \frac{(1-\nu)}{E}\frac{d}{dR}(RS_\theta) - \frac{\nu}{E}\frac{d}{dR}(RS_R)$$

or

$$(1-\nu)\frac{d}{dR}(RS_\theta) - \nu\frac{d}{dR}(RS_R) - S_R + 2\nu S_\theta = 0$$

From Figure 6.3 the force equilibrium gives

$$2S_\phi R\frac{d\theta}{2}dR + 2S_\theta R\frac{d\phi}{2}dR + \left(S_R + \frac{dS_R}{dR}dR\right)((R+dR)d\theta(R+dR)d\phi$$

$$- S_R dR(Rd\theta)(Rd\phi) = 0$$

With $S_\phi = S_\phi$, $d\theta = d\phi$

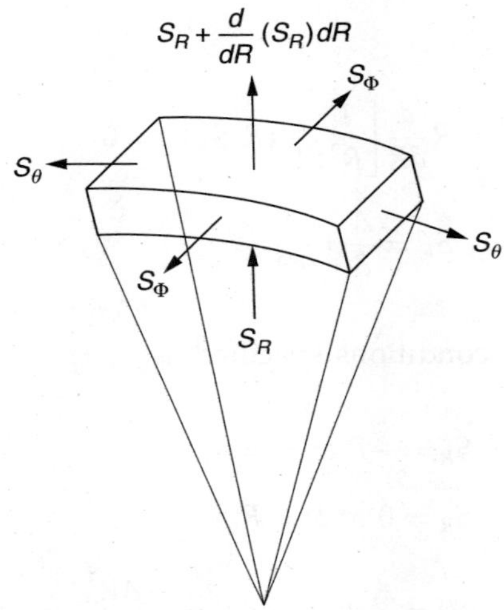

***Figure* 6.3** Equilibrium of a hemispherical element.

$$2RS_\theta dRd\theta^2 + \left(S_R + \frac{dS_R}{dR}dR\right)(R^2 + 2RDR)d\theta^2) - S_R R^2 d\theta^2 = 0$$

$$2RS_\theta dRd\theta^2 + S_R R^2 d\theta^2 + 2RS_R dRd\theta^2 + R^2\frac{dS_R}{dR}dRd\theta^2 + 2R\frac{dS_R}{dR}\frac{(dR)^2}{d\theta^2}$$
$$- S_R R^2 d\theta^2 = 0$$

$$2S_\theta = -2RS_R - R\frac{dS_R}{dR}$$
$$= \frac{-1}{R}\frac{d}{dR}(R^2 S_R)$$
$$S_\theta = -\frac{1}{2R}\frac{d}{dR}(R^2 S_R)$$

Substituting, we have

$$-\frac{(1-\nu)}{2}\frac{d}{dR}\left[\frac{1}{R^2}\frac{d}{dR}(R2S_R)\right] - \nu\frac{d}{dR}(RS_R) - S_R - \frac{\nu}{R}\frac{d}{dR}(R^2 S_R) = 0$$

which gives

$$R\frac{d}{dR}\left[\frac{1}{R^2}\frac{d}{dR}\left(R^3S_R\right)\right]=0$$

$$S_R=\frac{A}{3}+\frac{B}{R^3} \tag{6.6}$$

With the boundary conditions specified as

$$S_R=-P \text{ at } R=R_\text{i}$$

$$S_R=0 \text{ at } R=R_\text{o} \tag{6.7}$$

$$0=\frac{A}{3}+\frac{B}{R_\text{o}^3} \text{ or } B=\frac{-AR_\text{o}^3}{3}$$

Therefore

$$S_R=\frac{A}{3}\left(1-\frac{R_\text{o}^3}{R_\text{i}^3}\right)$$

and

$$-P=\frac{A}{3}\left(1-\frac{R_\text{o}^3}{R_\text{i}^3}\right)$$

This gives

$$A=\frac{3R_\text{i}^3P}{R_\text{o}^3-R_\text{i}^3};\ B=\frac{-AR_\text{o}^3}{3}$$

$$S_R=\frac{PR_\text{i}^3}{R_\text{o}^3-R_\text{i}^3}\left(1-\frac{R_\text{o}^3}{R^3}\right) \tag{6.8}$$

and

$$S_\theta=S_\phi=\frac{PR_\text{i}}{R_\text{o}^3-R_\text{i}^3}\left(1+\frac{R_\text{o}^3}{2R^3}\right) \tag{6.9}$$

6.3 ASME equation for hemispherical heads

ASME Section VIII Division 1 provides the following equation for internal pressure.[1] This is a compromise between a thin-shell equation and "exact" equation.

The design thickness of a hemispherical head is given by

$$t = \frac{PR}{25E - 0.2P} \tag{6.10}$$

where R is the inside radius, S is the allowable shear, and $E =$ is the joint efficiency.

6.4 Example problem 1

A hemispherical head having an inside radius of 380 mm is subjected to an internal pressure of 28 Megapascals (MPa). This allowable stress is 160 MPa. What is the required thickness using the shell theory and "exact" theory, and the ASME equation (assume joint efficiency, $E = 1$)?

6.4.1 Thin-shell theory

From Eq. (6.1) membrane stress

$$S = \frac{PR}{2t}$$

or

$$t = \frac{PR}{2s} = \frac{28 \times 380}{320} = 33.25 \text{ mm}$$

taking the radius as the inside radius.

6.4.2 "Exact" theory

Using Eq. (6.9)

$$S = (S_\theta = S_\phi) = \frac{PR_i^3}{R_o^3 - R_i^3}\left(1 + \frac{R_o^3}{2R_i^3}\right)$$

which simplifies to

$$R_o = \sqrt{\frac{2(S+P)R_i^3}{2S - P}}$$

Thus

$$R_o = \sqrt{\frac{2(160 + 28) \times 380^3}{2(160) - 28}} = 413.4 \text{ mm}$$

Therefore

$$t = R_o - R_i = 413.4 - 380 = 33.4 \text{ mm}$$

Therefore the assumption of this shell theory is valid here.

6.4.3 ASME equation (assuming E = 1)

Using Eq. (6.10)

$$t = \frac{28 \times 380}{2(160)(1) - 0.2(28)} = 33.8 \text{ mm}$$

The ASME estimate is conservative in this case.

6.5 ASME design equation for ellipsoidal heads

For an internal pressure P, the thickness t of the ellipsoidal head is given by

$$t = \frac{PDK}{2SE - 0.2P} \qquad (6.11)[1]$$

where D = diameter of the shell to which the head is attached, E = joint efficiency, S = allowable stress, and K = stress intensity factor.

K is given by the following expression:

$$K = \frac{1}{6}\left[2 + \left(\frac{a}{b}\right)^2\right] \qquad (6.12)$$

where a and b are the semi-major and semi-minor axes of the ellipse.

6.6 ASME equation for torispherical heads

For an internal pressure P, the thickness of the torispherical head is given by

$$t = \frac{PLM}{2SE - 0.2P} \qquad (6.13)[1]$$

where L = spherical cross radius, S = allowable stress, E = joint efficiency, and M = shear intensity factor. M, the stress intensity factor

$$M = \frac{1}{4}\left(3 + \sqrt{\frac{L}{r}}\right) \tag{6.14}$$

where r is the knuckle radius. The special case when the knuckle radius is 6 percent of the spherical crown radius, or $r = 0.06L$ is known as ASME head.

For the ASME head, $M = 1.77$ (from Eq. (6.14)) and the thickness t is then given by

$$t = \frac{0.885PL}{SE - 0.1P} \tag{6.15}$$

It turns out that for large ratios of R/t, the knuckle region of the head is prone to buckling under internal pressure. Based on plastic analysis,[1] the following expression is used for t:

$$\ln\frac{t}{l} = -1.26177 - 4.55246\left(\frac{r}{D}\right) + 28.9133\left(\frac{r}{D}\right)^2 + \left[0.66299 - 2.24709\left(\frac{r}{D}\right)^2\right] +$$

$$\left[0.66299 - 2.24709\left(\frac{r}{D}\right) + 15.62899\left(\frac{r}{D}\right)^2\right]\ln\frac{P}{S} + .$$

$$\left[0.26879 \times 10^{-4} - 0.44262\left(\frac{r}{D}\right) + 1.88783\left(\frac{r}{D}\right)^2\right]\left(\ln\frac{P}{S}\right)^2$$

where L = crown radius, r = knuckle radius, D = diameter of the shell to which the head is attached, and S = allowable stress.

6.7 Example problem 2

What is the required thickness of a torispherical head attached to a shell of diameter 6 mm, to have a crown radius of 6 mm and a knuckle radius of 360 mm? (ASME head $r/L = 0.06$). The allowable stress is 120 MPa and the internal pressure is 345 KPa.

6.7.1 Solution for ASME head using Eq. (6.15)

$$t = \frac{0.885PL}{SE - 0.1P}$$

assuming $E = 1$. With $S = 120$ MPa, we have

$$t = \frac{0.885(0.345)6}{(120)(1) - (0.1)(0.345)} = 0.0153 \text{ m} = 15.3\text{mm}$$

The thickness is small compared to the diameter of the head and should be checked for buckling at the knuckle region of the head.

We also have

$$\frac{r}{D} = \frac{0.36}{6} = 0.06$$

$$\frac{P}{S} = \frac{0.345}{120} = 0.002875$$

$$\ln\frac{P}{S} = -5.8517$$

$$\left(\ln\frac{P}{S}\right)^2 = 34.2424 \tag{6.16}$$

We have using Eq. (6.16)

$$\ln\left(\frac{t}{l}\right) = -1.26177 - 4.55246(0.06) + 28.93316(0.06)^2 +$$

$$[0.66299 - 2.4709(0.06) + 15.68299(0.06)^2](-5.8517)$$

$$+[0.26879 \times 10^{-4} - 0.44262(0.06) + 1.88783(0.06)^2](34.2424)$$

$$= -1.26177 - 0.27315 + 0.10416 - 5.8517[0.66299 - 0.13483 + 0.05646]$$

$$+34.824[0.26879 \times 10^{-4} - 0.02656 + 0.0680] = -5.53897$$

This gives $t/L = 0.00393$, or $t = (60000)(0.00393) = 23.6$ mm. Hence a minimum thickness 23.6 mm is required. The design is therefore dictated by stability of the knuckle region of the head.

6.8 ASME design equations for conical heads

ASME Code Section VIII Division I provides the following equation for thickness t of conical heads subjected to an internal pressure P.[1] With α as the semi-apex angle of the cone

$$t = \frac{PD}{2\cos\alpha(SE - 0.6P)} \tag{6.17}$$

where D is the inside diameter of cone measured perpendicular to longitudinal axis, S is the allowable stress, and E is the joint efficiency

6.9 ASME design equations for toriconical heads

A toriconical head is a blend of conical and torispherical heads. Accordingly, the thickness, t_c in the cone region is calculated using conical head equations and that in the head transition section is calculated using torispherical head equations.

Referring to Figure 6.4 for the conical region we have, using Eq. (6.17),

$$t_c = \frac{PD_1}{2\cos\alpha(SE - 0.6P)} \tag{6.18}$$

and for the torispherical region using Eq. (6.13)

$$t_k = \frac{PLM}{2SE - 0.2P} \tag{6.19}$$

where

$$L = \frac{D_1}{2\cos\alpha}$$

and

$$M = \frac{1}{4}\left(3 + \sqrt{\frac{L}{r}}\right)$$

from Eq. (6.14)

A pressure vessel designer generally has flexibility in selecting head geometry. Most common is of course the torispherical head, which is characterized by inside diameter, crown radius, and knuckle radius. The designer selects a head configuration that minimizes the total cost of the plate material and its formation.

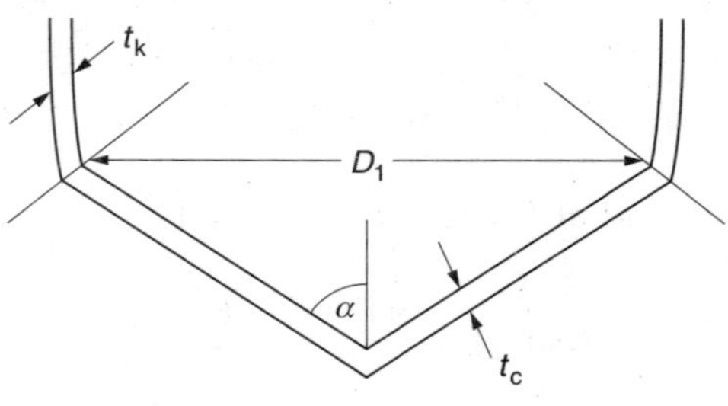

Figure 6.4 Toriconical head.

6.10 Flat heads and covers

Flat heads or covers are used widely as closures to pressure vessels. They are either integrally formed with the shell, or may be attached by bolts. Figure 6.5 shows some typical designs of covers.

6.10.1 Case 1

A simply supported circular plate of radius R and thickness t subjected to uniform pressure P. The deflection at the center of this plate is a maximum and this value is given by[2,3]

$$\delta_{\text{max}} = \frac{5+\nu}{1+\nu} \times \frac{PR^4}{64D} \tag{6.20}$$

where

$$D = \frac{Et^3}{12(1-\nu^2)} \tag{6.21}$$

where t = plate thickness.

The stress is a maximum at the bottom surface[2,3]

$$(S_r)_{\text{max}} = (S_\theta)_{\text{max}} = \frac{3(3+\nu)}{8}\frac{PR^2}{t^2} \tag{6.22}$$

6.10.2 Case 2

A circular plate is clamped around outer periphery and subjected to uniform pressure P. The maximum deflection occurs at the center of the plate where the value is[2,3]

$$\delta_{\text{max}} = \frac{PR^4}{64D} \tag{6.23}$$

The maximum radial and tangential stresses are given by[2,3]

$$(S_{\text{r}})_{\text{max}} = \frac{3PR^2}{4t^2} \tag{6.24}$$

occurring at the edge and at the top surface, and

$$(S_\theta)_{\text{max}} = \frac{3(1+\nu)}{8}\,\frac{PR^2}{t^2} \tag{6.25}$$

occurring at the center and the top surface of the plate.

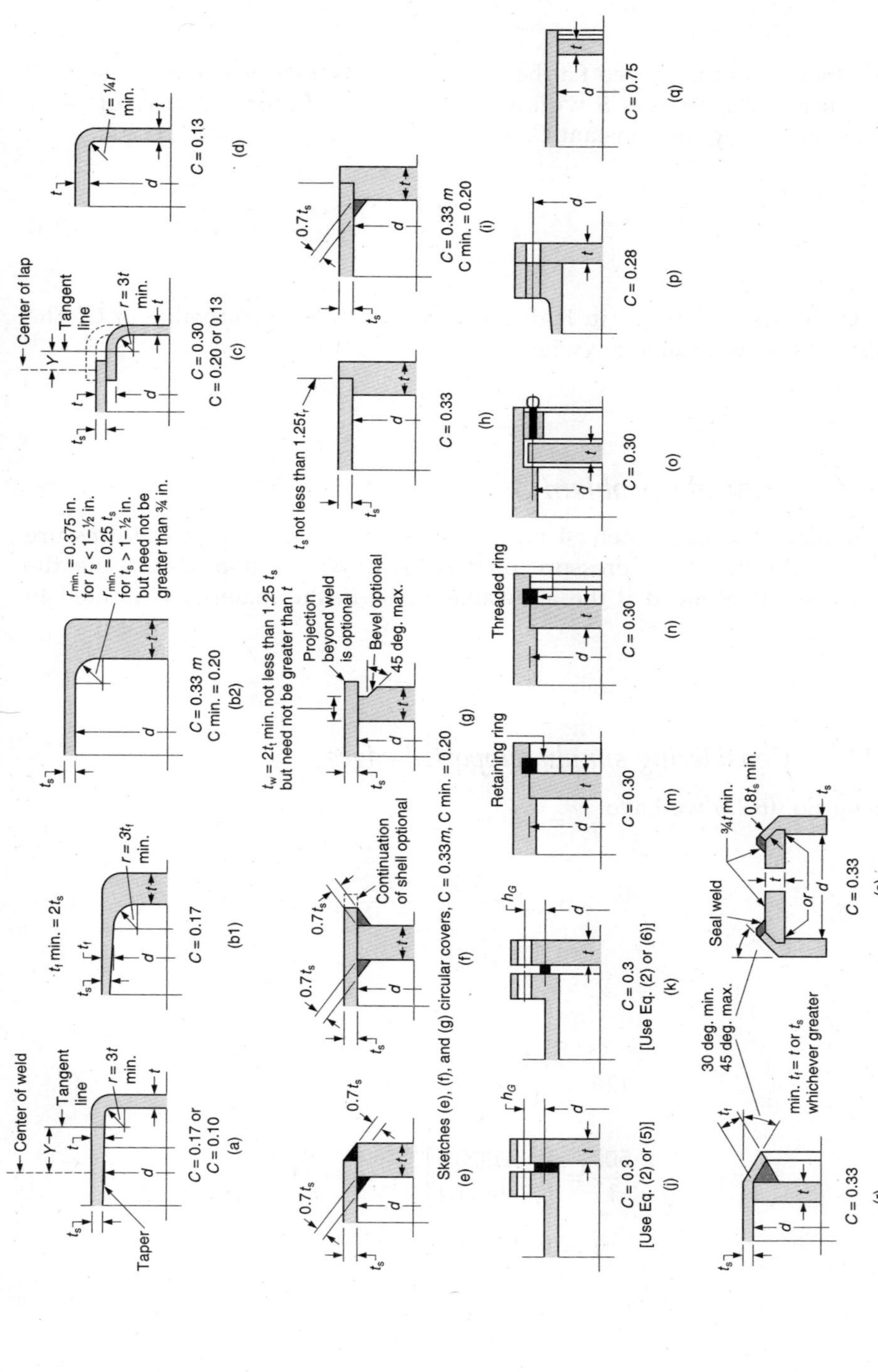

Figure 6.5 Cover plate designs. (Modified from ASME Boiler and Pressure Vessel Code, ASME, New York.)

6.11 ASME equation for unstayed flat heads and covers

The thickness of unstayed flat heads and covers subjected to a pressure P, and an allowable stress S with a joint efficiency E, for a variety of cases characterized by the constant C, is given by

$$t = d\sqrt{\frac{CP}{SE}} \tag{6.26}$$[1]

The cases are shown in Figure 6.5 each with a typical value of C. The value of C could range anywhere from 0.10 to 0.33.

6.12 Example problem 3

A circular plate of diameter 1 m, forms the cover for a cylindrical pressure vessel subjected to a pressure of 0.04 MPa. We wish to determine the thickness of the head if the allowable stress in the material is limited to 120 MPa.

6.12.1 Considering simply supported edges

Using Eq. (6.22) we have

$$S_{\max} = \frac{3(3+\nu)}{8}\frac{PR^2}{t^2}$$

or

$$120 = \frac{9.9}{8}(0.04)\left(\frac{500}{t}\right)^2$$

$$\frac{500}{t} = \left[\frac{120(8)}{(9.9)(0.04)}\right]^{(1/2)} = 49.24$$

$$t = 10.16 \text{ mm}$$

6.12.2 *Considering clamped edges*

$$S_r|_{max} > S_\theta|_{max}$$

$$\frac{500}{t} = \left[\frac{120(4)}{0.12}\right]^{(1/2)} = 63.25$$

$$t = 7.91 \text{ mm}$$

6.12.3 *Considering unstayed plates and covers*

See Figure 6.5. We have from Eq. (6.26)

$$t = d\sqrt{\frac{CP}{SE}}$$

where $C = 0.10$–0.33 depending on construction, d = diameter of the head, P = design pressure, S = allowable tensile stress, and E = butt weld joint efficiency. Assuming $E = 1$, then $S = 120$ MPa.

For $C = 0.10$

$$t = 1000\sqrt{\frac{0.10(0.04)}{(1)(120)}} = 5.77 \text{ mm}$$

For $C = 0.33$

$$t = 1000\sqrt{\frac{0.33(0.04)}{(1)(120)}} = 10.49 \text{ mm}$$

References

1. American Society of Mechanical Engineers, Boiler and Pressure Vessel Code, ASME, New York.
2. Timoshenko, S.P., and Woinowsky-Kreiger, S., *Theory of Plates and Shells*, 2nd ed., McGraw-Hill, 1959.
3. Roark, R.J. and Young, W.C., *Formulas for Stress and Strain*, 5th ed., McGraw-Hill, New York, 1975.

chapter seven

Design of nozzles and openings

Contents

7.1 Introduction

Openings in pressure vessels in the regions of shells or heads are required to serve the following purposes:

- Manways for letting personnel in and out of the vessel to perform routine maintenance and repair
- Holes for draining or cleaning the vessel
- Hand hole openings for inspecting the vessel from outside
- Nozzles attached to pipes to convey the working fluid inside and outside of the vessel

For all openings, however, nozzles may not be necessary. In some cases we have nozzles and piping that are attached to the openings, while in other cases there could be a manway cover plate or a handhole cover plate that is welded or attached by bolts to the pad area of the opening. Nozzles or openings may be subjected to internal or external pressure, along with attachment loads coming from equipment and piping due to differential thermal expansion and other sources.

The design of openings and nozzles is based on two considerations:

1. Primary membrane stress in the vessel must be within the limits set by allowable tensile stress.
2. Peak stresses should be kept within acceptable limits to ensure satisfactory fatigue life.

Because of removal of material at the location of the holes, there is a general weakening of the shell. The amount of weakening is of course dependent on the diameter of the hole, the number of holes, and how far the holes are spaced from one another. One of the ways the weakening is accommodated for is by introducing material either by weld deposits or by forging. The aspects of stress intensification as well as reinforcement will be addressed in this chapter.

7.2 Stress concentration about a circular hole

The stress concentration caused by a hole in a plate due to uniaxial tension or biaxial tension will be considered. The biaxial tension would correspond to a cylindrical shell or a spherical shell subject to internal pressure. For the case of a cylindrical shell, the biaxiality is 2:1 corresponding to hoop and longitudinal strain; for the case of a spherical shell, the biaxiality ratio is 1:1.

The radial and tangential components of the stresses at a distance r from the center of the hole of radius a (see Figure 7.1) are given by

$$\sigma_r = \frac{\sigma}{2}\left(1 - \frac{a^2}{r^2}\right) + \frac{\sigma}{2}\left(1 + \frac{3a^4}{r^4} - \frac{4a^2}{r^2}\right)\cos 2\theta \tag{7.1}$$

and

$$\sigma_t = \frac{\sigma}{2}\left(1 + \frac{a^2}{r^2}\right) - \frac{\sigma}{2}\left(1 + \frac{3a^4}{r^4}\cos 2\theta\right) \tag{7.2}$$

and

$$\sigma_{rt} = -\frac{\sigma}{2}\left(1 - \frac{3a^4}{r^4} + \frac{2a^2}{r^2}\right)\sin 2\theta \tag{7.3}$$

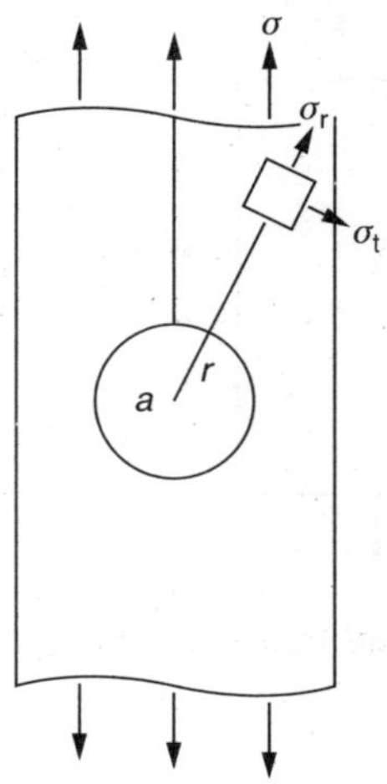

Figure 7.1 Radial and tangential stresses in a uniaxially loaded plate with a hole.

The maximum value of the tangential component σ_t occurs at the edge of the hole where $r = a$ and $\theta = \pi$:

$$\sigma_t|_{\max} = \frac{\sigma}{2}\left(2 + \frac{a^2}{r^2} + \frac{3a^4}{r^4}\right) = 3\sigma \tag{7.4}$$

We can see that as r gets very large, the maximum σ_t approaches the far-field stress, σ. This gives the following stress distribution:

$$\begin{aligned} \sigma_t|_{\max} &= 3\sigma \text{ at } r = a \\ &1.15\sigma \text{ at } r = 2a \\ &1.07\sigma \text{ at } r = 3a \end{aligned} \tag{7.5}$$

7.3 *Cylindrical shell with a circular hole under internal pressure*

If we indicate the hoop stress by σ, then the longitudinal stress will be $\sigma/2$ and the maximum stress will be the combined effect of the combination of the hoop and longitudinal stresses. Then $\sigma_{\text{hoop}} = \sigma$, which will be acting at $\theta = \pi/2$ and $\sigma_{\text{long}} = \sigma/2$, acting at $\theta = 0$.

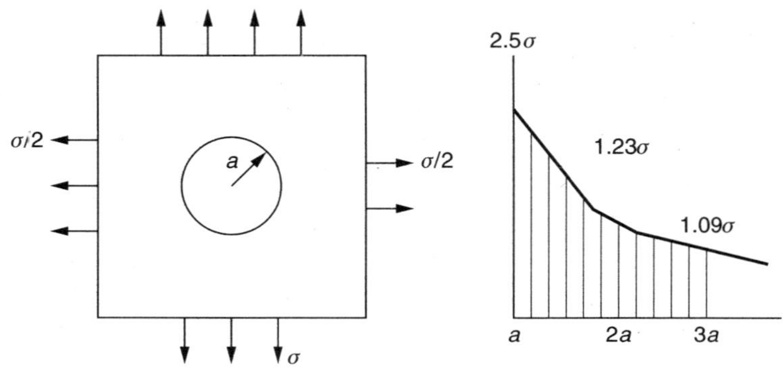

Figure *7.2* Tangential stress due to a hole in an internally pressurized cylindrical shell.

Adding the contribution of these two cases, we have

$$\begin{aligned}\sigma_t|_{\max} &= \frac{\sigma}{2}\left(1+\frac{a^2}{r^2}\right)+\frac{\sigma}{2}\left(1+\frac{3a^4}{r^4}\right)+\frac{\sigma}{4}\left(1+\frac{a^2}{r^2}\right)-\frac{\sigma}{4}\left(1+\frac{3a^4}{r^4}\right)\\ &= \frac{\sigma}{4}\left(4+\frac{3a^2}{r^2}+\frac{3a^4}{r^4}\right)\end{aligned} \tag{7.6}$$

This gives the following stress distributions (see Figure 7.2):

$$\left.\begin{aligned}\sigma_t &= 2.5\alpha \text{ at } r = a\\ &1.23\sigma \text{ at } r = 2a\\ &1.09\sigma \text{ at } r = 3a\end{aligned}\right\} \tag{7.7}$$

7.4 *Spherical shell with a circular hole under internal pressure*

Here both the hoop stress and the meridional stress would be σ:

$$\sigma_t|_{\max} = \sigma\left(1+\frac{a^2}{r^2}\right) \tag{7.8}$$

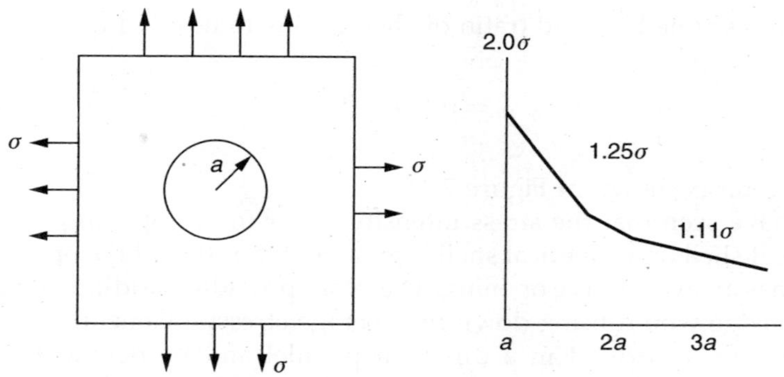

Figure 7.3 Tangential stress due to a hole in an internally pressurized spherical shell.

This gives the following stress distributions (see Figure 7.3):

$$\sigma_t = 2\sigma \text{ at } r = a$$
$$1.25\sigma \text{ at } r = 2a$$
$$1.11\sigma \text{ at } r = 3a \quad (7.9)$$

7.5 Reinforcement of openings

The philosophy is based on providing additional material in the region of the opening by thickening the shell or adding a pad material. The additional material is deemed effective in carrying the higher loads. On most vessels, is provided on the outside of the vessel. In some vessels, the reinforcement appears inside, while in others both inside and outside regions are reinforced. On many vessels, however, the arrangement is such that no reinforcement can be placed on the inside because of interfering components.

The placement of this additional material is important. We note that at a distance of $r = 2a$ for all three situations, namely a plate under tension, a cylindrical shell under internal pressure, and a spherical shell under internal pressure, the stresses die out sufficiently. So this distance is generally taken at the boundary limit for the effective reinforcement to the vessel surface. This is indicated in Figure 7.4

For the reinforcement limit perpendicular to the vessel wall in the direction of the nozzle axis is based on the direction of the nozzle. Here we use the theory of a beam on an elastic foundation[1] where the length is given by

$$L = \frac{1}{\beta} = \frac{\sqrt{at_{\rm n}}}{1.285} \quad (7.10)$$

For an assumed $\frac{a}{t_n} = 10$ (ratio of shell radius to nozzle radius)

$$L = 0.25a = 2.5t_n \tag{7.11}$$

This limit is shown in Figure 7.4.

We have seen that the stress intensity factor for an opening in either a cylindrical shell or a spherical shell is greater at the edge of the opening and diminishes away from the opening. Therefore providing additional material near the edge would bring down the average stresses. Generally the limits of reinforcement extend in a direction parallel and perpendicular to the surface of the shell, and are based on the assumption that the added reinforcement adequately compensates for the loss of structural integrity as a result of material removal at the opening.

The limit parallel to the surface of the shell is typically set as the larger of two quantities: (a) $t_s + t_n + 0.5d$ and (b) d; where t_s and t_n are the shell and the nozzle thickness, respectively, and d is the diameter of the opening. If we consider d as the controlling dimension then the stress intensity factor at a distance d from the center of the hole in a cylindrical shell is 1.23 and that for the spherical shell is 1.25. It is judged that with the added reinforcement the nominal stress could be reduced close to that of a solid shell.

The limit normal to the surface of the shell measured inward or outward is typically set as the smaller of (a) $2.5t_s$ or (b) $2.5t_n$.[2]

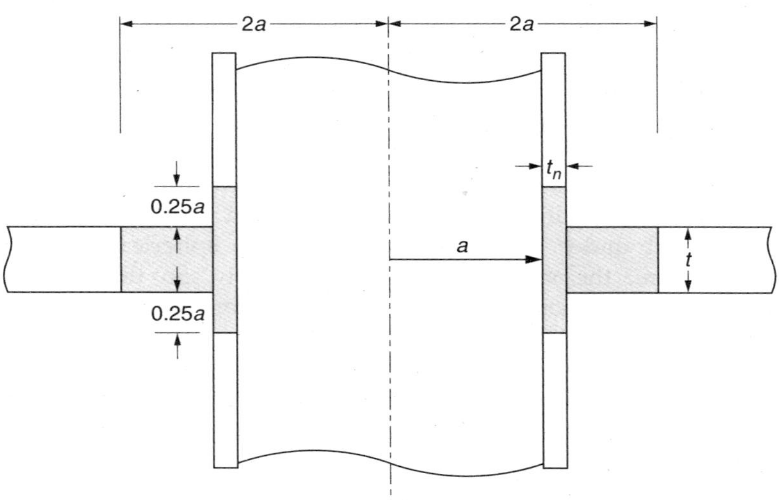

Figure 7.4 Circumferential and transversal reinforcement extents.

7.5.1 *Reinforcement example problem*

Determine the reinforcement requirements for a 300 mm diameter opening in a cylindrical pressure vessel 1 m in diameter subjected to an internal pressure of 5 MPa. The shell and the nozzle allowable stress is 120 MPa. The shell and nozzle thickness are 25 mm and 32 mm, respectively.

Minimum required shell thickness is given by

$$t_{rs} = \frac{PR_s}{S - 0.6P} = \frac{5(500)}{120 - 0.6(5)} = 21.4 \text{ mm} \tag{7.12}$$

Minimum required nozzle thickness is given by

$$t_{rn} = \frac{PR_n}{S - 0.6P} = \frac{5(150)}{120 - 0.6(5)} = 6.4 \text{ mm} \tag{7.13}$$

The limit parallel to the surface of the shell is the larger of two quantities, (a) $t_s + t_n + 0.5d = 32 + 25 + 150 = 207$ mm, and (b) $d = 300$ mm. Parallel to the shell the limit is therefore 300 mm.

The limit normal to the surface of the shell is the smaller of (a) 2.5 t_s or (b) 2.5 t_n. Therefore the limit is 2.5 (25) = 62.5 mm.

The reinforcement scheme is shown in Figure 7.5. The reinforcement area required is

$$A_r = dt_{rs} = (300)(21.4) = 6420 \text{ mm}^2 \tag{7.14}$$

The reinforcement area available in the shell (up to a distance d), A_1 is given by

$$A_1 = (2d - d)(t_s - t_{sr}) = 300(25 - 21.4) = 1080 \text{ mm}^2 \tag{7.15}$$

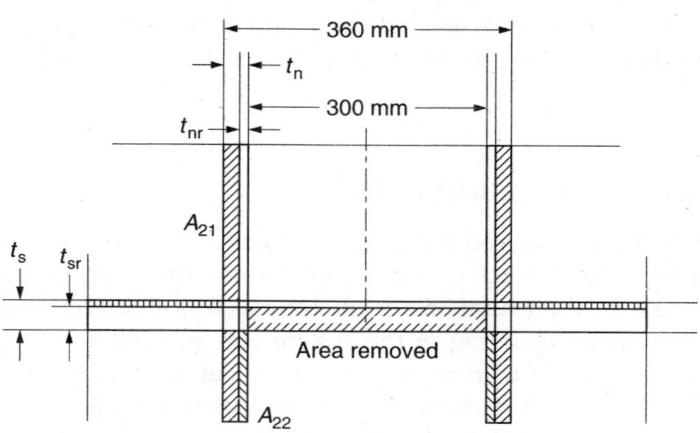

Figure 7.5 Reinforcement scheme for the example problem.

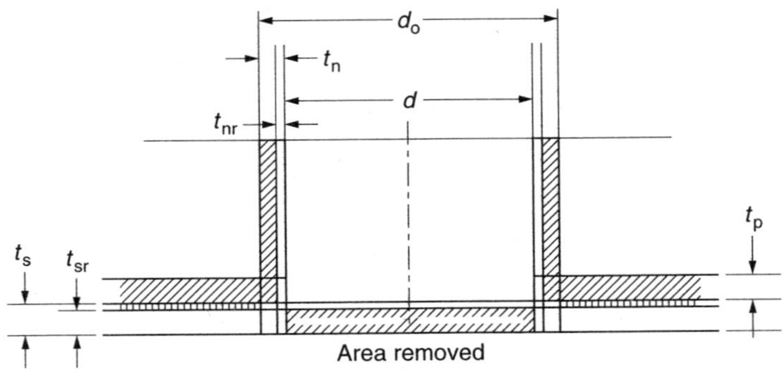

Figure 7.6 Alternative reinforcement scheme.

The reinforcement area available in nozzle wall is available in two parts A_{21} and A_{22},

$$A_{21} = 2(2.5t_s)(t_s - t_{rn}) = 2(2.5)(25)(32 - 6.4) = 3200 \text{ mm}^2 \quad (7.16)$$

$$A_{22} = 2(2.5t_s)(t_s) = 2(2.5)(25)(32) = 4000 \text{ mm}^2 \quad (7.17)$$

The total area available for reinforcement A_t is then

$$A_t = A_1 + A_{21} + A_{22} = 1080 + 3200 + 4000 = 8280 \text{ mm}^2 \quad (7.18)$$

The available reinforcement therefore exceeds the reinforcement required (Eq. (7.14)) and the method is acceptable.

An alternative form of reinforcement is shown in Figure 7.6 for a flush nozzle. The additional area is $2dt_p$, where we have provided an extra thickness t_p on the shell for reinforcement purposes.

One of the main disadvantages of this reinforcement method is that it gives no information on stresses and these can vary significantly from one design to another resulting in differing performances, especially for cyclic loadings.

7.6 Nozzles in pressure vessels

Two cases will be considered here, namely that for spherical vessels and for cylindrical vessels. Leckie and Penny[3] treat the case of nozzles in spherical vessels in their analysis of an intersecting cylinder and sphere. The maximum stress occurring in the sphere was presented in a graphical form. Both flush and protruding nozzles were considered in the analysis. The stress concentration factors have been calculated in terms of the maximum stress in the sphere by neglecting the bending stresses. In its most basic form, a radial nozzle in a spherical vessel is defined by four

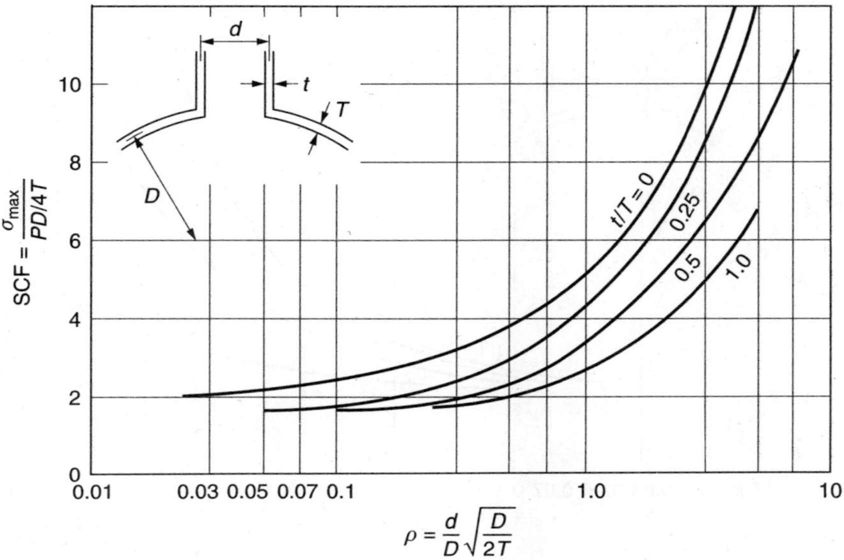

***Figure** 7.7* Stress concentration factors (SCFs) for a nozzle in a spherical shell.

geometric variables. These are the nozzle and sphere diameters d and D, and the corresponding thicknesses t and T. The stress concentration factors can be expected to increase as the ratio d/D increases (keeping the same D/T, of course). This is evident from Figure 7.7. However because of the thin-shell assumptions concerning the joining of the nozzle and the sphere, the solutions can only be expected to describe the gross structural behavior.

The cylinder/cylinder geometry (for nozzles in cylindrical vessels) is much more difficult to analyze than the nozzle sphere, which can be treated as an axisymmetric structure. To obtain a suitable stress concentration factor for a nozzle in a cylindrical vessel, an approximate axisymmetric model is sometimes used. A popular approximation used is where the equivalent sphere has twice the diameter of the shell. The general trend of the experimental results is shown in Figure 7.8, which is for a flush nozzle in a cylindrical shell.

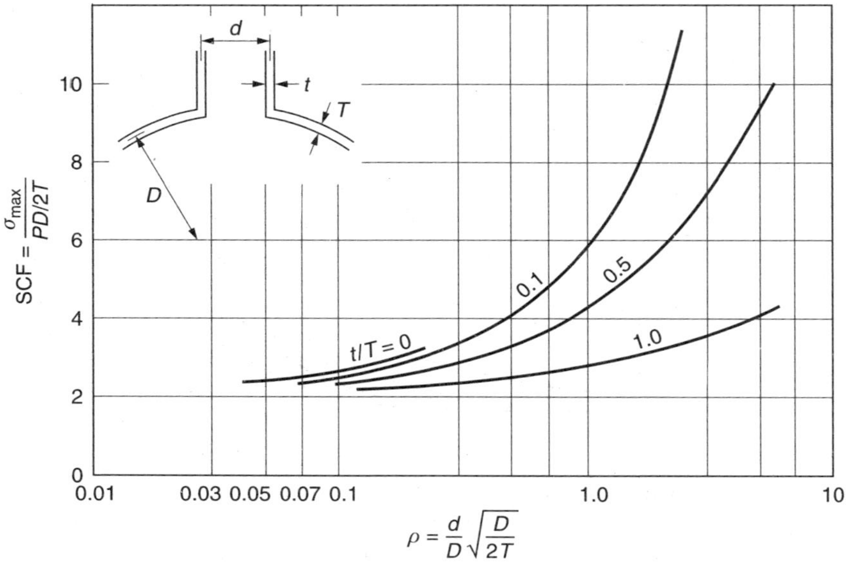

Figure *7.8* Stress concentration factors (SCFs) for a nozzle in a cylindrical shell (approximate).

References

1. Hetenyi, M., *Beams on Elastic Foundation*, University of Michigan Press, Ann Arbor, 1958.
2. American Society of Mechanical Engineers, Boiler and Pressure Vessel Code, ASME, New York.
3. Leckie, F.A., and Penny, R.K., *Weld. Counc. Res. Bull.* No. 90, 1963.

chapter eight

Fatigue assessment of pressure vessels

Contents

8.1 Introduction

Fatigue has been recognized as a major failure mode in pressure vessels, and specific rules for its prevention appear in design codes. Stated simply, fatigue failure is caused by the cyclic action of loads and thermal conditions. In many design situations, the expected number of cycles is in millions and for all practical purposes can be considered as infinite. Accordingly, the concept of endurance limit has been employed in a number of design rules.

Endurance limit is the stress that can be applied for an infinite number of cycles without producing failure. However, the typical number of stress cycles rarely exceeds 100,000 and frequently only a few thousand. Therefore, fatigue analysis requires somewhat more involved concepts than just the endurance limit.

Fatigue refers to the behavior of material under repeated loads, which is distinct from the behavior under monotonically applied loading. There is a progressive localized permanent deformation under fluctuating loads that culminates in cracks and complete fracture after a sufficient number of cycles. The fatigue process itself occurs over a period of time. However, failure may occur suddenly and without prior warning, in which case the damage mechanisms may have been operating since loading was first introduced. This period of time is often referred to as the usage period. The fatigue process appears to initiate from local areas that have high stresses. These highly stressed regions are due to abrupt changes in geometry leading to high stress concentrations, due to temperature differentials, imperfections, or the presence of residual stress. The failure takes place when the crack after repeated cycling grows to a point at which the material can no longer withstand the loads and a complete separation occurs. Metallurgical defects such as a void or an inclusion often act as sites for fatigue crack initiation. The fatigue process consists of crack initiation, crack propagation, and eventual fracture. Another way to look at the process is to postulate it in terms of initiation of microcracks, coalescence of these microcracks into macrocracks, followed by growth to unstable fracture.

Fatigue has been classified as one of high cycle and low cycle. High cycle fatigue involves very little plastic action. The low cycle fatigue failure involves a few thousand cycles and involves strains in excess of yield strain. Fatigue damage in the low cycle has been found to be related to plastic strain and fatigue curves for use in this region should be based on strain ranges. For the high cycle fatigue cases, the stress ranges can be used. The procedure of using strain amplitude as a function of number of cycles forms the cornerstone of fatigue analysis. The design curve is based on strain-controlled data. The best-fit curves were reduced by a factor of 2 on stress and 20 on cycles to account for environment, size effect, and scatter of data. The basic elements of the fatigue evaluation in pressure vessels rests on the use of maximum shear theory of failure (Tresca criterion), with the assumptions of linear elastic behavior along with the use of Miner's rule for estimating the cumulative effect of stress cycles of varying amplitude.

Fatigue failure typically occurs at structural discontinuities which give rise to stress concentration. The stress concentration factors are generally based on theoretical analysis involving statically applied loads. These are directly applicable to fatigue analysis only when the nominal stress multiplied by the stress concentration factor is below the material yield strength. When it exceeds the yield strength, there is a redistribution of stress and strain. For sharp geometries, using values of stress concentrations obtained elastically leads to underprediction of fatigue lives when compared with the actual test data. In fact under no circumstances does a value of more than 5 need be applied, and it is observed that the value of these factors do not vary with the magnitude of cyclic strain and associated fatigue life. In the design of pressure vessels, the stress concentration – which is really the strain concentration – is limited to 5 and for most discontinuities such as grooves and fillets no more than a value of 4 is used.

8.2 Exemption from fatigue analysis

These rules apply to the situations where the possibility of fatigue failure is remote. The motivation for this effort stems from the fact that most pressure vessels are subjected to limited number of pressure and temperature cycles during their lifetime, and therefore considerable design effort could be saved by defining conditions that do not require a fatigue evaluation to be performed, an approach first proposed by Langer.[1] In this procedure, the designer needs to know the pressure fluctuations, and an estimate of temperature differences between different points in the vessel. The six rules are as follows:

1. The specified number of pressure cycling does not exceed the number of cycles on the design fatigue curve, corresponding to the stress amplitude of $3S_m$ (typically twice the yield strength).
2. The specified full range of pressure fluctuation during normal operation does not exceed the quantity $S_a/S3m$ times the design pressure, where S_a is the value obtained from the applicable design fatigue curve for total number of significant pressure fluctuations. Significant pressure fluctuations are those for which the excursion exceeds the quantity (design pressure $S_a/S3m$) where S is the value of S_a for 10^6 cycles.
3. The temperature difference between any adjacent points (points separated less than $2\sqrt{Rt}$ from each other, R = mean radius, t = thickness) during normal operation and during startup and shutdown operation does not exceed $S_a/2E_\alpha$, where S_a is the alternating stress for the specified number of startup and shutdown cycles.
4. The temperature difference between two adjacent points of the vessel does not change during normal operation by more than the quantity $S_a/2E_\alpha$, where S_a is the value obtained from the applicable design fatigue curve for the total specified number of temperature fluctuations. Significant temperature range exceeds $S_a/2E_\alpha$, where S is the value of S_a for 10^6 cycles.
5. For components fabricated from dissimilar materials, the total algebraic range of temperature fluctuation does not exceed $S_a/2(E_1\alpha_1 - E_2\alpha_2)$, where S is the value obtained from the design fatigue curve for total specified number of significant temperature fluctuations. In this case, the significant temperature fluctuation is one for which the total excursion exceeds the quantity $S_a/2(E_1\alpha_1 - E_2\alpha_2)$, where S is the value of S_a for 10^6 cycles.
6. The specified full range of mechanical loads does not result in load stresses whose range exceeds S_a, a value obtained from the design fatigue curve for total specified number of significant load fluctuations. A load fluctuation is considered significant if the total excursion of load stress exceeds the value of S_a for 10^6 cycles.

The fatigue exemption rules outlined above are based on a set of assumptions, some of which are conservative and some of which are not so conservative. It is believed, however, that conservative ones outweigh the nonconservative ones.

A stress concentration factor of 2 is assumed at a point where the nominal stress is $3S_m$. This leads to a peak stress of $6S_m$ due to pressure and is thus quite conservative. The calculated stress due to temperature difference ΔT between two points does not exceed $2E\alpha\Delta T$, which is amplified to $4E\alpha\Delta T$ due to the assumed stress concentration factor of 2. This value bounds all the applicable cases of linear thermal gradient, thermal shock, and gross thermal mismatch, and the assumption is indeed a very conservative one. Finally the concept of adjacent points where the two points in the pressure vessel that are separated by a distance more than $2\sqrt{Rt}$ from each other, allows for sufficient flexibility to produce a significant reduction in thermal stress.

8.3 S–N curves

Typical strain–life curves are obtained under load or strain control tests on smooth specimens. Here S refers to the applied stress, usually taken as the alternating stress, S_a and N is the number of cycles to failure (complete fracture or separation). Constant amplitude S–N curves are usually plotted on a semi-log or log–log coordinates and often contains data points with scatter as shown in Figure 8.1. Some of the fatigue curves show a curve that slopes continuously downward with the number of cycles. Other fatigue curves contain a discontinuity or a knee in the S–N curve as shown in Figure 8.2. The number of cycles causing fatigue failure depends on the strain level incurred during each cycle. The fatigue curves obtained from experimental data provide the variation of cyclic strain amplitude (elastic plus plastic strain) with the number of cycles to failure. The critical strain amplitude, at which a material may be cycled indefinitely without

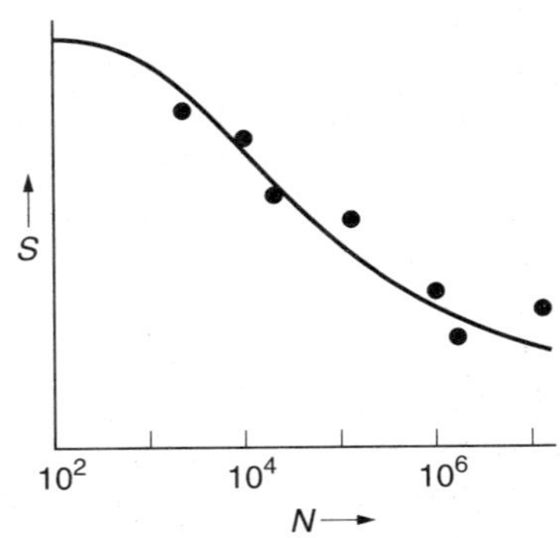

Figure 8.1 Typical S–N curve showing data scatter.

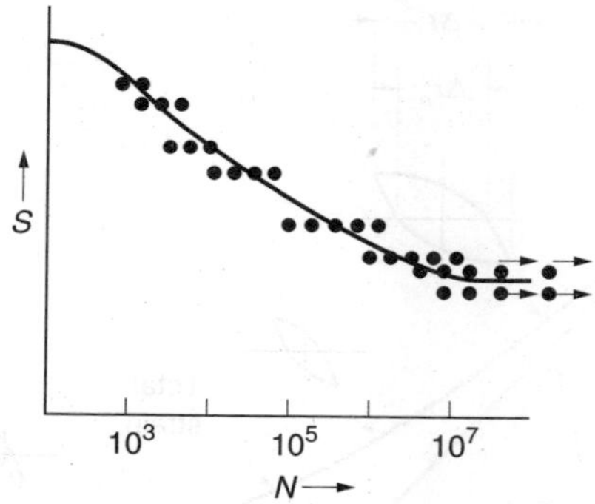

Figure 8.2 S–N curve with discontinuity (knee).

producing failure is the endurance limit strain. Fatigue data in the form of S–N curve are generally obtained at room temperature. The strain range obtained from the test is converted to nominal stress range by multiplying by the modulus of elasticity. Half of the stress range is the alternating stress, S_a which appears along the ordinate of the S–N curve.

The endurance limit, as mentioned earlier, is the cyclic stress amplitude which will not cause fatigue failure regardless of the number of cycles of load application. The exact endurance limit is never really found, since no test specimen is cycled to infinite number of cycles. For pressure vessels the components are normally not cycled beyond 10^6 cycles and therefore the fatigue limit is defined as the stress amplitude that will cause fatigue failure in 10^6 cycles. In most situations, fatigue limit and the endurance limit are used interchangeably and obviously signify stress amplitude that produces fatigue failure in 10^6 cycles.

8.4 Local strain approach to fatigue

Based on the initiation or formation of microcracks, fatigue life can be estimated from the knowledge of local strain–time history at a notch in a component and the unnotched strain–life fatigue properties, along with a cumulative damage model. Strain–life fatigue data are obtained from tests using small polished unnotched axial fatigue specimens under constant amplitude reversed cycles of strain. Steady-state hysteresis loops can predominate through most of the fatigue life, and this can be reduced to elastic and plastic strain ranges. A typical strain–life curve on a log–log scale is shown in Figure 8.3. At a given value of N (the number of cycles to failure) the total strain range is the sum of elastic and plastic strain ranges. Both the elastic and plastic strain curves are straight lines in the log–log plot, having a slope of b and c, respectively. The following equation is used for the total strain range, $\Delta\varepsilon$.[2]

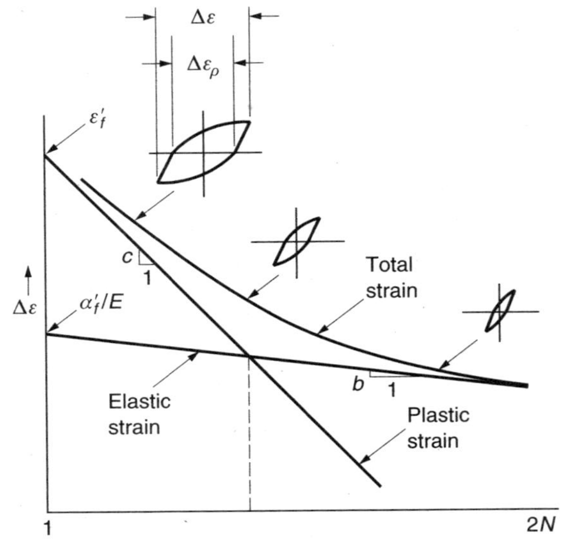

Figure 8.3 Typical strain–life curve.

$$\frac{\Delta\varepsilon}{2} = \frac{\Delta\varepsilon_e}{2} + \frac{\Delta\varepsilon_p}{2} = \frac{\sigma_f'}{E}(2N)^b + \varepsilon_f'(2N)^c \tag{8.1}$$

Here $\Delta\varepsilon/2$ = total strain amplitude, $\Delta\varepsilon_e/2$ = elastic strain amplitude = $\sigma/(2E)$, $\Delta\varepsilon_p/2$ = plastic strain amplitude = $\Delta\varepsilon/2 - \Delta\varepsilon_e/2 = \Delta\varepsilon/2 - \sigma/(2E)$, $\varepsilon_f{}'$ = fatigue ductility coefficient (related to fracture strain in a monotonic tensile test), and $\sigma_f{}'$ = fatigue strength coefficient (related to fracture stress in a monotonic tensile test).

To determine the local strain range, the cyclic stress strain curve is used, since under the influence of cyclic loads, the material will soon approach the stable cyclic condition. The cyclic stress–strain curve is computed as

$$\frac{\Delta\sigma}{2} = K'(\Delta\varepsilon_p/2)^{n'} \tag{8.2}$$

where $\Delta\sigma$ = true stress range, and $\Delta\varepsilon_p$ = true plastic strain range.

If the nominal stress range is indicated by ΔS, then the nominal strain $\Delta e = \Delta S/E$. Then according to Neuber's rule[3]

$$\Delta\sigma \times \Delta\varepsilon = \Delta S \times \Delta e = \frac{(\Delta S)^2}{E} \tag{8.3}$$

From Eqs. (8.2) and (8.3) we have

$$\Delta\varepsilon = \frac{\Delta\sigma}{E} + \left(\frac{\Delta\sigma}{2^{1-n'}K'}\right)^{\frac{1}{n'}} = \frac{(\Delta S)^2}{E^2\Delta\varepsilon} + \left[\frac{(\Delta S)^2}{2^{1-n'}K'E\Delta\varepsilon}\right] \tag{8.4}$$

The number of cycles to failure can be obtained by first finding the total strain and then using the strain–life relationship.

8.5 Design fatigue curves

In the design codes such as the ASME Boiler and Pressure Vessel Code,[4] a safety factor of 2 on the stress amplitude and 20 on cycles to failure is used. If the stress calculations are made using a material having a modulus of elasticity E_{calc} which is different from the modulus used for the design curve, E_{curve}, then the computed alternating stress, S'_a should be corrected to obtain the alternating stress used in the fatigue curve as

$$S_a = S'_a \frac{E_{curve}}{E_{calc}} \tag{8.5}$$

Next the appropriate fatigue curve is employed. The fatigue curve needed by the designer is one which shows stress, S_a, versus cycles to failure, N, and which contains sufficient safety factors to give safe allowable design stress for a given number of operating cycles, or conversely, allowable operating cycles for a given value of calculated stress. The ordinate value is S_a, as modified by Eq. (8.5). The abscissa for the fatigue curve is the number of cycles, N, allowed by the alternating stress intensity, S_a.

The design fatigue curves for carbon steel to account for the adjustment for maximum mean stress is shown in Figure 8.4. Similar curves for austenitic steels and nickel–chrome steels are shown in Figure 8.5. Both these curves have been obtained from the ASME Boiler and Pressure Vessel Code.[4] For 2024-T4 aluminum, the fatigue curve is shown in Figure 8.6.[5]

8.6 Cumulative damage

Pressure vessels are subjected to a variety of cyclic loadings which could be either consecutively or concurrently applied. By consecutively applied loads we mean that in a component a stress range of magnitude ΔS_1 is applied for n_1 cycles followed by another stress range of magnitude ΔS_2 is applied for n_2 cycles, and so on. When, however, such stress ranges are applied concurrently, the stress ranges get added up.

For consecutive application of cyclic loadings, experimental results indicate that there is a fatigue usage, or the fraction of fatigue life expended as measured by the ratios n_1/N_1 or n_2/N_2; where n_1, n_2 are the number of

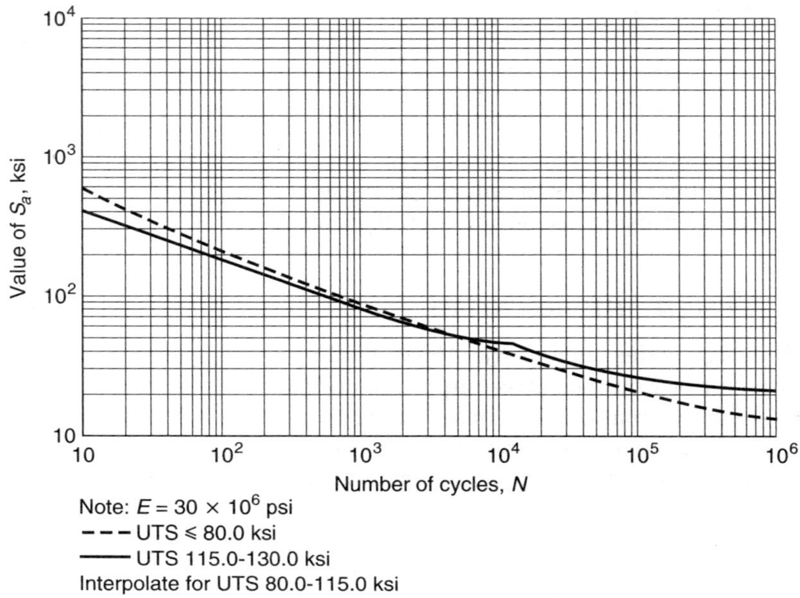

Figure 8.4 Design fatigue curve for carbon steel (ASME).
(Modified from ASME Boiler and Pressure Vessel Code, ASME, New York.)

actual cycles of stress ranges ΔS_1 and ΔS_2 and N_1, N_2 are the number of cycles to failure at these stress ranges. These ratios are called partial usage factors. Adding up all these individual partial usage factors, we obtain the cumulative fatigue usage factor, which if less than unity ensures that the component is safe from fatigue failure due to cyclic loadings. Usage factor must be below unity to satisfy ASME code compliance. When there are two

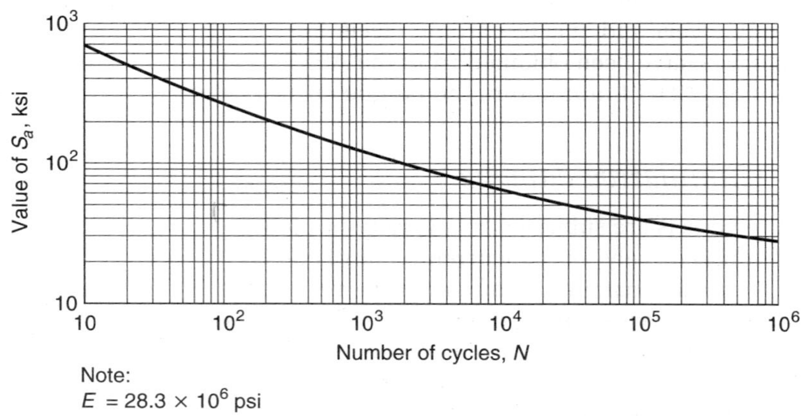

Figure 8.5 Design fatigue curve for stainless steel (ASME).
(Modified from ASME Boiler and Pressure Vessel Code, ASME, New York.)

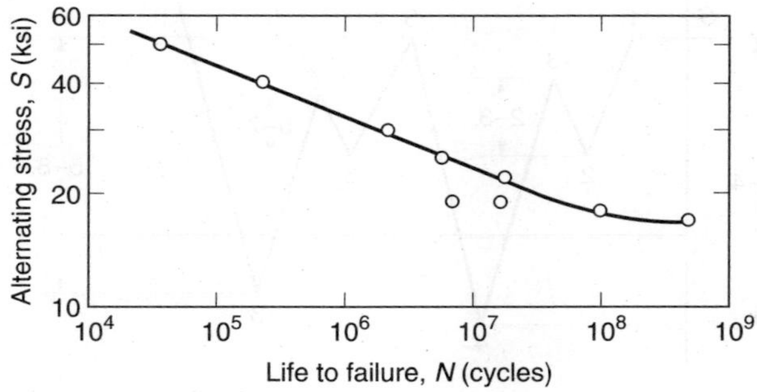

Figure 8.6 Fatigue curve for 2024-T4 aluminum.

or more stress cycles, which produce significant stresses, the cumulative effect shall be evaluated as follows:

- Designate the specified number of times each cycle will be repeated during the life of the vessel by n_1, n_2, n_3, etc. In determining n_1, n_2, n_3, etc, consideration shall be given to the superposition of cycles of various origins which produce a total stress difference range greater than the stress difference ranges of the individual cycles.
- Calculate the cumulative usage factor U by means of a linear damage relationship known as Miner's rule.[6] It is assumed that if n_1 cycles would produce failure at a stress level S_1 then n_1 cycles at the same stress level would use up the fraction $\frac{n_1}{N}$ of the total life. Failure occurs when the cumulative usage factor, which is the sum of $\frac{n_1}{N_1} + \frac{n_2}{N_2} + \frac{n_3}{N_3} + \ldots$ is equal to 1.0.
- Other hypotheses for estimating cumulative fatigue damage exist and some have the potential for predicting fatigue damage more accurately than the linear damage model. However, this is by far the simplest method and has been backed up by tests[7] that show that the linear assumption is quite good when the cycles of large and small stress magnitude are fairly evenly distributed throughout the life of the vessel. It is obvious that when the sequence of the stress cycles is known in considerable detail, better accuracy can be obtained.

8.7 Cycle counting

Let us have a stress time history characterized peaks and valleys as shown in Figure 8.7. The history is first rearranged to start with the highest peak. Starting from the highest peak we go down to the next reversal. We next

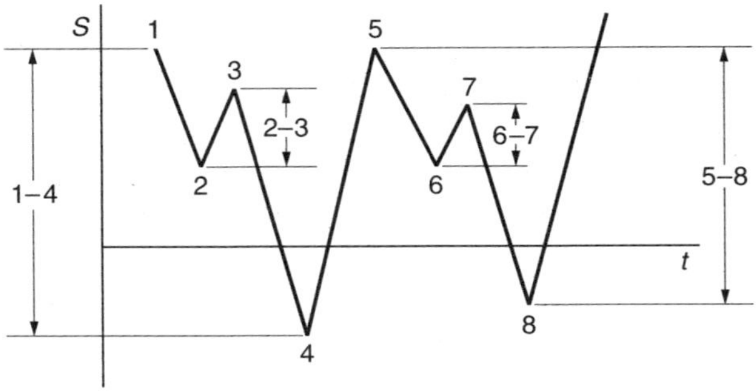

Figure 8.7 Rainflow cycle counting.

proceed horizontally to the next downward range. If there is no range going down from the level of the valley that we have stopped, we go upward to the next reversal. Looking at Figure 8.7, the result of the count may be 1–4; 3–2; 5–8; and 7–6.

This procedure is called rainflow counting because the lines going horizontally from a reversal to a succeeding range resembles rain flowing down a pagoda roof when the history of peaks and valleys is turned around 90 degrees.

Generally speaking this would be the case when the number of cycles associated with each peak or valley is completely exhausted in the counting process. In actual stress time histories there may be cycles left over which should be used for the next count involving the next lesser stress range and so on.

8.8 Fatigue evaluation procedure

The fatigue evaluation is based on performing a detailed stress analysis, extracting the principal stress, stress differences, and alternating stress intensities. The allowable number of cycles is obtained from the design fatigue curve and incremental fatigue usages are obtained by determining the ratios of calculated number of cycles. The incremental usages are then added to obtain the cumulative fatigue usage.

The fatigue evaluation methodology is based on the minimum shear stress theory of failure. This consists of finding the amplitude (one half of the range) through which the maximum shear stress fluctuates. This is obtained in the ASME Code procedure by using stress difference and stress intensities (twice the maximum shear stress).

At each point on the vessel at a given time, there are three principal stresses, σ_1, σ_2, σ_3, and three stress differences $S_{1,2}$, $S_{2,3}$, and $S_{3,1}$. The principal stresses are calculated from the six components of the stress tensor. The directions of the principal stresses may change during the cycle but each principal stress retains its identity as it rotates. The stress differences are calculated as a function of time for the complete cycle and the largest absolute magnitude of any stress difference at any time is determined. The alternating stress intensity S_a is one half of this quantity. The specific process used is outlined as follows.[1]

For a given loading the six stress components are evaluated first. This procedure is employed for all the other loadings. If the stress cycle consists of a stress excursion that occurs between two instants (or load conditions) #1 and #2, then the stress components for the first load condition is obtained as

$$\sigma_{xx}^1,\ \sigma_{yy}^1,\ \sigma_{zz}^1,\ \tau_{xy}^1,\ \tau_{yz}^1,\ \tau_{zx}^1$$

The stress components for the second load condition is obtained as

$$\sigma_{xx}^2, \sigma_{yy}^2, \sigma_{zz}^2, \tau_{xy}^2, \tau_{yz}^2, \tau_{zx}^2$$

The principal stresses associated with the first and the second load condition is calculated as

$$\sigma_1^1, \sigma_2^1, \sigma_3^1; \quad \sigma_1^2, \sigma_2^2, \sigma_3^2$$

The corresponding stress differences for the first load condition are calculated as

$$S_{1,2}^1, S_{2,3}^1, S_{3,1}^1$$

and those for the second load condition as

$$S_{1,2}^2, S_{2,3}^2, S_{3,1}^2$$

where

$$S_{1,2}^1 = \sigma_1^1 - \sigma_2^1;\ S_{2,3}^1 = \sigma_2^1 - \sigma_3^1;\ S_{3,1}^1 = \sigma_3^1 - \sigma_1^1 \tag{8.6}$$

$$S_{1,2}^2 = \sigma_1^2 - \sigma_2^2;\ S_{2,3}^2 = \sigma_2^2 - \sigma_3^2;\ S_{3,1}^2 = \sigma_3^2 - \sigma_1^2 \tag{8.7}$$

The maximum stress range at two instants during the cycle is chosen in such a way that the maximum stress range is obtained as

$$S_{1,2}^{\mathrm{r}} = S_{1,2}^1 - S_{1,2}^2;\ S_{2,3}^{\mathrm{r}} = S_{2,3}^1 - S_{2,3}^2;\ S_{3,1}^{\mathrm{r}} = S_{3,1}^1 - S_{3,1}^2 \tag{8.8}$$

The largest absolute magnitude of $S_{1,2}^r$; $S_{2,3}^r$ and $S_{3,1}^r$ is designated S_r. Half of S_r is the stress amplitude, which is used for component fatigue evaluation. However this procedure implicitly assumes that the principal axes do not rotate between the instants. For the general multiaxial loading in which the principal axes are no longer fixed but indeed moving then the fatigue evaluation is modified as follows.[1] If the cycle occurs between the two instants 1 and 2 as before, then the pseudo-stress components are evaluated as

$$\delta\sigma_{xx} = \sigma_{xx}^1 - \sigma_{xx}^2 \tag{8.9}$$

$$\delta\sigma_{yy} = \sigma_{yy}^1 - \sigma_{yy}^2 \tag{8.10}$$

$$\delta\sigma_{zz} = \sigma_{zz}^1 - \sigma_{zz}^2 \tag{8.11}$$

$$\delta\tau_{xy} = \tau_{xy}^1 - \tau_{xy}^2 \tag{8.12}$$

$$\delta\tau_{yz} = \tau_{yz}^1 - \tau_{yz}^2 \tag{8.13}$$

$$\delta\tau_{zx} = \tau_{zx}^1 - \tau_{zx}^2 \tag{8.14}$$

The principal values of these pseudo-stress components are then calculated. Let us call them δS_1, δS_2 and δS_3. The stress differences are designated as

$$\delta S_{1,2} = \delta S_1 - \delta S_2 \tag{8.15}$$

$$\delta S_{2,3} = \delta S_2 - \delta S_3 \tag{8.16}$$

$$\delta S_{3,1} = \delta S_3 - \delta S_1 \tag{8.17}$$

The largest absolute magnitude of δS_1, δS_2 and δS_3 is designated as S_r. Half of S_r is the stress amplitude which is used for entering into the design fatigue curve. It may be necessary to investigate many pairs of instants to find the maximum value of S_r.

8.9 *Example of fatigue evaluation*

A carbon steel pressure vessel is subjected to 400 pressure cycles with a stress range of 1430 MPa followed by 600 high-temperature cycles with a stress range of 590 MPa and then 300 low-temperature cycles with a stress range of 440 MPa. Is a fatigue failure likely?

For the stress range of 1430 MPa – that is an alternating stress of 715 MPa – the number of cycles to failure from Figure 8.4 is obtained as 600. For the stress range of 590 MPa (alternating stress of 295 MPa) the number of cycles to failure is 7000; that for the stress range of 440 MPa (alternating stress of 220 MPa), the number of cycles to failure is 20,000.

We have $n_1 = 400, N_1 = 600; n_2 = 600, N_2 = 7{,}000$; and $n_3 = 3000, N_3 = 20{,}000$. Therefore the cumulative fatigue usage, U, is given by

$$U = \sum_{i=1}^{3} \frac{n_i}{N_i} = \frac{n_1}{N_1} + \frac{n_2}{N_2} + \frac{n_3}{N_3} = \frac{400}{600} + \frac{600}{7000} + \frac{3000}{20000} = 0.90$$

The cumulative fatigue usage is 0.9, a value that is less than unity. Therefore fatigue failure is unlikely to occur.

References

1. Langer, B.F., Design of pressure vessels for low cycle fatigue, *J. Basic Eng.*, September 389–402, 1962.
2. Fuchs, H.O., and Stephens, R.I., *Metal Fatigue in Engineering*, Wiley, New York, 1980.
3. Neuber, H., Theory of stress concentration for shear strained prismatical bodies with arbitrary nonlinear stress strain laws, *J. Appl. Mech.*, 28, 544, 1961.
4. American Society of Mechanical Engineers, Boiler and Pressure Vessel Code, ASME, New York.
5. Osgood, C.C., *Fatigue Design*, 2nd ed., Pergamon Press, Oxford, 1982.
6. Miner, M.A. Cumulative damage in fatigue, *J. Appl. Mech.*, 12, A-159, 1945.
7. Baldwin, E.E., Sokol, G.J., and Coffin, Jr., L.B., Cyclic strain fatigue studies on AISI Type 347 stainless steel, *Am. Soc. Test. Mat. Proc.*, 57, 567, 1957.

Problems

1. A stainless steel pressure vessel is subjected to 2000 pressure cycles where the stress alternates between zero and 200 MPa. This is followed by 10,000 cycles of thermal stresses that alternate between zero and 800 MPa. Determine the cumulative fatigue usage.
2. A carbon steel pressure vessel is subjected to 1000 pressure cycles at a temperature where the elastic modulus may be taken as 190 MPa. The stress varies from zero to 400 MPa. This is followed by 8000 thermal cycles where the stress varies from –100 MPa to 500 MPa. Assume $E = 180$ GPa for the thermal cycles. Is the vessel safe from fatigue failure?
3. A pressure vessel made from carbon steel is subjected to a number of transients wherein the stress ranges are listed in decreasing order. The alternating stresses and the corresponding number of operating cycles are indicated in the table below. Determine the cumulative fatigue usage factor.

Transient range	Alternating stress: MPa	Number of cycles
1	400	10
2	360	50
3	250	500
4	200	5000
5	150	8000

4. Consider a stress discontinuity in the vessel of Problem 3 that gives rise to an effective stress concentration factor of 1.5 at which the nominal stress ranges from the table of Problem 3 is applicable. Determine the cumulative fatigue usage factor.

chapter nine

Bolted flange connections

Contents

9.1 Introduction

Bolted flange connections perform a very important structural role in the closure of flanges in a pressure vessel. Their importance stems from two important functions: (a) maintenance of the structural integrity of the connection itself, and (b) prevention of leakage through the gaskets preloaded by bolts.

A representative bolted flanged joint is shown in Figure 9.1. This is typically comprised of a flange ring and a tapered hub, which is welded to the pressure vessel. The flange is a seat for the gasket, and the cover (along with the gasket) is bolted to the flange by a number of bolts. The preload on the bolts is extremely important for the successful performance of the connection. The preload must be sufficiently large to seat the gasket and at the same time not excessive enough to crush it. The bolts should be designed to contain the pressure and for the preload required to prevent leakage through the gasket. The flange region should be designed to resist bending that occurs in the spacing between the bolt locations.

The gasket, which is really the focal point of the bolted flange connection, is subjected to compressive force by the bolts. The flange

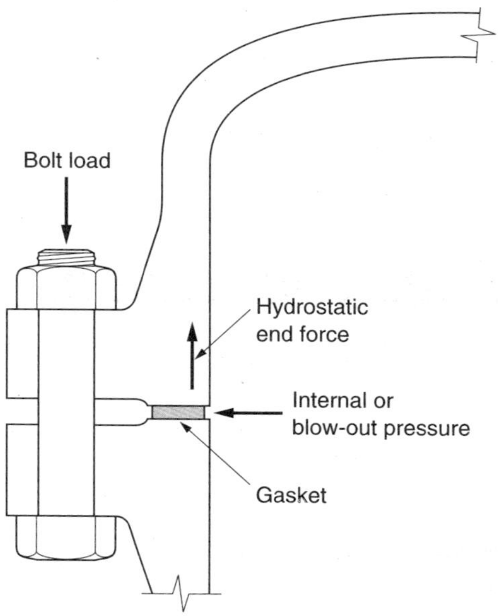

Figure 9.1 Bolted flange joint.

stiffness in conjunction with the bolt preload provides the necessary surface constraint and the compressive force to prevent the leakage of the fluid contained in the pressure vessel. The fluid pressure tends to reduce the bolt preload, which reduces gasket compression and tends to separate the flange faces. The gaskets are therefore required to expand to maintain the leak-proof boundary. Gaskets are made of nonmetallic materials with composite construction. The serrated surfaces of the flange faces help to maintain the leak-proof joint as the material expands to fill up the irregularities on the face of the flanges.

The mechanics of the bolted joint with gaskets is extremely complex to track analytically and experimental results are often used as bases for design. Some of the factors are determined experimentally.

9.2 *Gasket joint behavior*

The flange and bolts must meet two design requirements: (a) the gasket seating assembly condition, and (b) the operating condition. The design calculations use two gasket factors, the gasket seating stress, y, and the gasket factor at operating conditions, m. A higher gasket seating stress ensures better sealing performance. When the joint is in service, the fluid pressure load unloads the joint, resulting in a reduced gasket stress (a process illustrated in Figure 9.2). Under operating conditions it is advantageous to have a residual gasket stress greater than the fluid

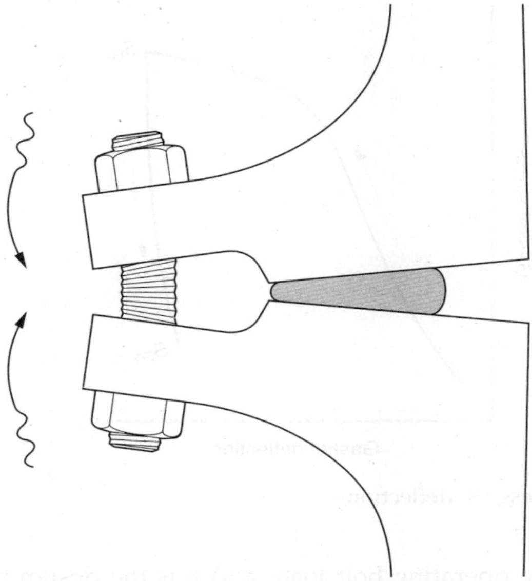

Figure 9.2 Flange bending.

pressure. For good sealing performance, the ASME Code recommends the residual stress at operating conditions be at least two to three times the contained pressure.[1] This is the so-called m factor. The relationship between the initial seating stress and the residual seating stress is indicated by the gasket stress vs. deflection plot shown in Figure 9.3. S_{G2} is the initial seating stress and S_{G1} is the operating gasket stress. During assembly, the gasket follows the nonlinear portion of the curve from zero to S_{G2}. When the operating pressure unloads the gasket, the gasket follows the unloading curve from points S_{G2} to S_{G1}.

In what follows we have used the notations of the ASME Boiler and Pressure Vessel Code, Section VIII, Division 1, Appendix 2.[1] The gasket joint structural characteristic is typically expressed in terms of the gasket seating stress, y, and the gasket factor, m. In the absence of fluid pressure, the required bolt load to seat the gasket is given by

$$W_{m2} = \pi b G y \tag{9.1}$$

Here W_{m2} is the gasket seating load, b is the effective gasket seating width, and G is the minimum gasket diameter. When the joint is pressurized by the fluid in the pressure vessel, the operating bolt load, W_{m1} is given by

$$W_{m1} = \frac{\pi}{4} G^2 p + (2b\pi G m p) \tag{9.2}$$

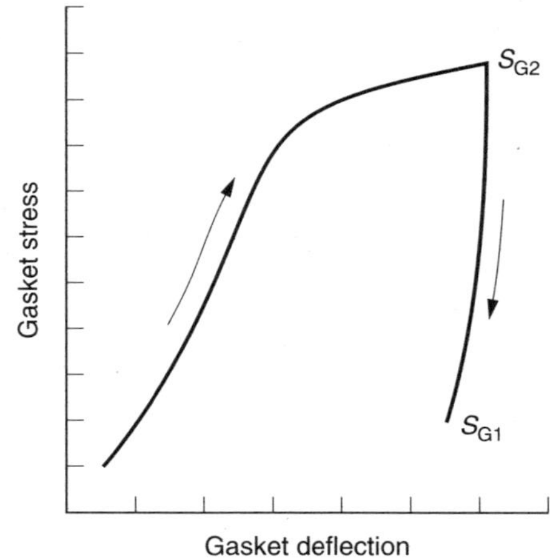

Figure 9.3 Gasket stress vs. deflection.

Here W_{m1} is the operating bolt load, and p is the design pressure. The first term in Eq. (9.2) is the joint end load due to fluid pressure. The second term is the joint contact load.

9.3 Design of bolts

The chosen bolt material should be compatible with the flange material. There must not be any chemical or galvanic action between the materials to preclude the possibility of thread seizure. The total minimum required cross-sectional area of the bolts should be the greater of the following areas:

$$\frac{W_{m2}}{S_a} \quad \text{and} \quad \frac{W_{m1}}{S_b}$$

Here S_a is the allowable bolt stress at room temperature, and S_b is the allowable bolt stress at design temperature.

9.4 Examples

9.4.1 Problem 1

The data for a bolted flange connection for a pressure vessel is given below. Should twelve 50-mm diameter bolts be adequate to ensure a leak-proof joint?

Design pressure $p = 17$ MPa
Gasket diameter $G = 382$ mm
Gasket width $b = 9$ mm

Gasket factor $m = 3$
Gasket seating stress $y = 69$ MPa
Bolt allowable stresses $S_a = S_b = 132$ MPa

Gasket loadings from Eqs. (9.1) and (9.2) are

$$W_{m2} = \pi(9)(382)(69) = 745.255 \text{ kN}$$

and

$$W_{m1} = \frac{\pi}{4}(382)^2(17) + 2(9)\pi(382)(3)(17) = (1948.344 + 1101.681)\,\text{kN} = 3050\,\text{kN}$$

The required bolt area is the greater of $(3050 \times 10^3)/132$ and $(745 \times 10^3)/132$, and is equal to 23,106 mm^2. Therefore, 12 bolts of diameter 50 mm give a bolt area of 23,562 mm^2 and should be adequate.

9.4.2 *Problem 2*

Calculate the required cover thickness for a 600-mm manway with the following design conditions:

Design pressure, $p = 4$ MPa
Flange material = SA 105
Allowable flange stress $S_a = 135$ MPa
Bolting material = SA 193 B7
Allowable bolt stress $S_b = 172$ MPa
Flange inside diameter = 590 mm
Flange outside diameter = 900 mm
Bolt circle diameter = 800 mm
The gasket is spiral-wound and graphite-filled with an outer ring.
Gasket outside diameter $G_{OD} = 690$ mm
Gasket inside diameter $G_{ID} = 630$ mm
Gasket properties: $m = 3.0$, $y = 69$ MPa

Figure 9.4 gives the schematic drawing of the cover flange. The gasket reaction location and the effective width are calculated as follows:

$N = 0.5\ (690–630) = 30$ mm
$b_o = N/2 = 15$ mm
Check whether the parameter b is greater than 6.3 mm.[1]
$b = 2.5\ (15)^{1/2} = 10$ mm (>6.3 mm)
Then $G_{OD} - 2b = 690–20 = 670$ mm (see Figure 9.4)
The moment arm, $h_G = 0.5\ (C–G) = 0.5\ (800–670) = 65$ mm (see Figure 9.4)

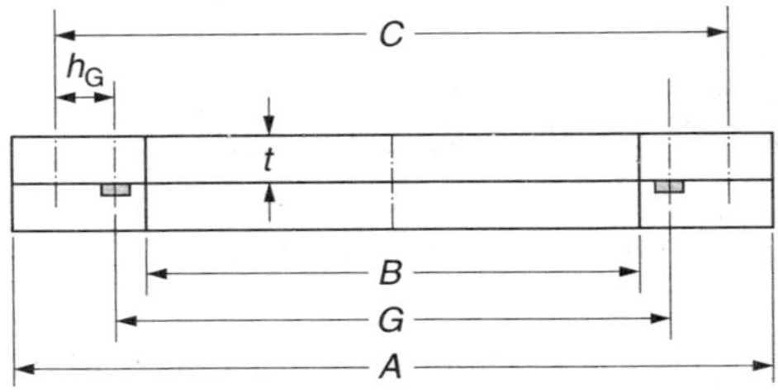

Figure 9.4 Gasketed blind flange.

W_{m2} = the minimum required bolt load for gasket seating
$= 3.14\ bGy$
$= 3.14\ (10)\ (670)\ 69$
$= 1452$ kN

W_{m1} = the minimum required bolt load during operation
$= 0.785G^2P + (2b)\ (3.14\ GmP)$
$= (0.785\ (670)^2 \times 4) + 20\ (3.14 \times 670 \times 3 \times 4)$
$= 1914$ kN

The required bolt areas are calculated as

A_{m1} = the required bolt area for operating case $= W_{m1}/S_b$
$= 1914 \times 10^3/172$
$= 11{,}128\ \text{mm}^2$

A_{m2} = the required bolt area for gasket seating $= W_{m2}/S_b$
$= 1452 \times 10^3/172$
$= 8442\ \text{mm}^2$

The required bolt area is the greater of A_{m1} and A_{m2}, and is 11,128 mm^2. For this application we will use 24 bolts of 39 mm major diameter for which the minor diameter area (root area) is 944 mm^2.[2] The actual bolt area selected is thus

$A_b = 24\ (944) = 22{,}656\ \text{mm}^2$
The operating load $W_{m1} = 1914$ kN

The modified equation for gasket seating to protect from overtightening and flange overloading is given by

$$W\ (\text{gasket seating}) = 0.5(A_b + A_{m2})\ S_a = 0.5(22{,}656 + 11{,}128)\ 135 = 2280\ \text{kN}$$

This is the load for the cold bolt-up case when there is no internal pressure applied.

The minimum required thickness of the cover plate is calculated by assuming the plate to be simply supported at the gasket load line (at diameter G in Figure 9.4) and loaded by a gasket load along with a uniform pressure load. This corresponds to an edge moment per unit length of the circumference of magnitude $Wh_G/(\pi G)$ giving rise to a stress at the center of the plate as

$$S = \frac{3(3+\upsilon)P}{32}\left(\frac{G}{t}\right)^2 + \frac{6}{\pi G}\left(\frac{Wh_G}{t^2}\right) \tag{9.3}$$

Using a value of 0.3 for the Poisson's ratio υ we have

$$S = 0.3P\left(\frac{G}{t}\right)^2 + 1.9\left(\frac{Wh_G}{Gt^2}\right) \tag{9.4}$$

Eq. (9.4) provides the following solution for t,

$$t = G\sqrt{\left[\frac{0.3P}{S} + \frac{1.9Wh_G}{SG^3}\right]} \tag{9.5}$$

A very similar equation appears in Reference 1.

The thickness of the cover is calculated for two cases and the larger dimension is used for design:

1. Case (a): operating condition
   $S = S_a = 135$ MPa, $P = 4$ MPa, $W = 1914$ kN. This gives

$$t = 800\sqrt{\left[\frac{0.3(4)}{135} + \frac{1.9(1914\times 10^3)65}{(135)800^3}\right]} = 89 \text{ mm} \tag{9.6}$$

2. Case (b): cold bolt up condition
   $S = 135$ MPa, $P = 0$, $W = 2280$ kN. This gives

$$t = 800\sqrt{\left[\frac{1.9(2280\times 10^3)65}{(135)800^3}\right]} = 51 \text{ mm} \tag{9.7}$$

The operating case thus governs the design and the minimum required thickness is 89 mm. With a corrosion allowance of 3 mm, the required thickness is 92 mm.

9.5 Closure

There is an important link between establishing a leak-free joint and meeting design requirements of an operating pressure vessel, particularly as it pertains to assembly. For the bolted flange connections that we have discussed, the gaskets are assumed to be entirely within the bolt circle. The m and y factors used in the design have been in existence since the 1940s. The ASME Boiler and Pressure Vessel Code lists the values of these factors; however, these are only suggested values and are not mandatory. Over the years these factors are being continuously revised to reflect further understanding of the joint performance. Recently research investigations on flanged joints have focused on tightness-based design. This concept is based on the understanding that all flange joints leak. The objective therefore is to design a flange joint that would ensure an acceptable leakage rate for the enclosed fluid.

References

1. American Society of Mechanical Engineers, Boiler and Pressure Vessel Code, ASME, New York.
2. American Society of Mechanical Engineers, Metric Screw Threads – M Profile, ASME B 1.13 M-1995, ASME, New York, 1995.

Problems

Calculate the required cover thickness for a test pressure vessel with the following design conditions:

Design pressure $p = 3.5$ MPa
Allowable flange stress $S_a = 130$ MPa
Bolting M 18×2.5 mm (stress area $= 175\ \text{mm}^2$) $= 8$ bolts
Allowable bolt stress $S_b = 275$ MPa
Bolt circle diameter $= 245$ mm
Gasket outside diameter $= 190$ mm
Gasket inside diameter $= 180$ mm
Gasket properties: $m = 3.0$, $y = 69$ MPa

chapter ten

Design of vessel supports

Contents

10.1 Introduction

The vessel support is intended to support the pressure vessel on the support base. The support has to be designed to withstand the dead weight and seismic loadings from the pressure vessel and to limit the heat flow from the vessel wall to the base. The pressure vessel support structure should be able to withstand the dead weight of the vessel and internals and the contained fluid without experiencing permanent deformation. The metal temperature of the pressure vessel is usually different to the ambient conditions during its installation. The differential displacement between the supports due to the temperature change should be considered in design. In a large number of cases the design of support requires adequacy to operate in a severe thermal environment during normal operation as well as to sustain some thermal transients. The other source of thermal loading arises from the thermal expansion of the piping attached to the vessel. The design must therefore consider the various combination of piping loads on the vessel to determine the most severe load combinations. In addition the vessel is also subject to mechanical loads due to the action of seismic accelerations on the attached piping. In large vessels containing liquids, the so-called "sloshing" effects must also be considered. Finally, loads due to handling during installation should be carefully considered in design.

The supports for pressure vessels can be of various types including lug support, support skirts, and saddle supports.

10.2 Lug support

This is a common means of support for vertical vessels that are mounted on I-beams. Such a support is shown in Figure 10.1. If the vessel is made of carbon steel, the lugs may be directly welded to the vessel. Bijlaard's classic assessment of local stresses in shells due to loadings on an attached lug is particularly noteworthy.[1] That analysis forms the basis for the Welding Research Council Bulletin 107[2] which has been used extensively for the design of lug attachments to pressure vessels. The method consists of determining the stresses in the vicinity of a support lug of height $2C_1$ and width $2C_2$ as shown in Figure 10.2. The maximum primary plus secondary stress in the shell wall is given as a combination of direct stress due to the thrust, W, bending stress due to longitudinal moment, M_L, bending stress due to circumferential moment, M_C, and the torsional shear stress due to the twisting moment, M_T, with appropriate coefficients. The earlier work by Bijlaard[1] involves representing M_L, M_C, and M_T, by double Fourier series, which enables one to obtain the stresses and deformations in the form of the series. In other words, the series is capable of representing a load with dimensions in both the circumferential, φ, and the longitudinal, x, directions:

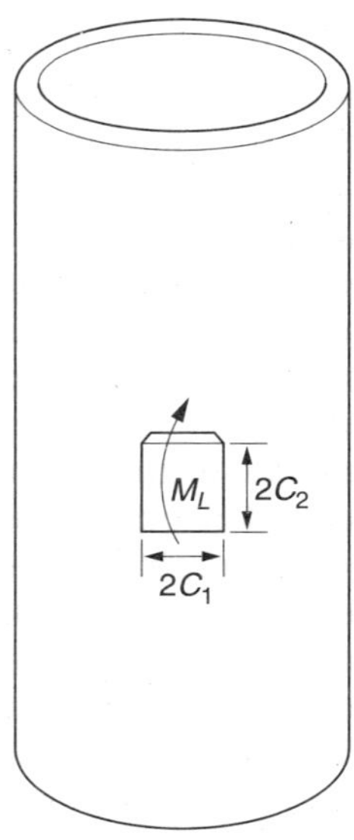

Figure 10.1 Lug supports.

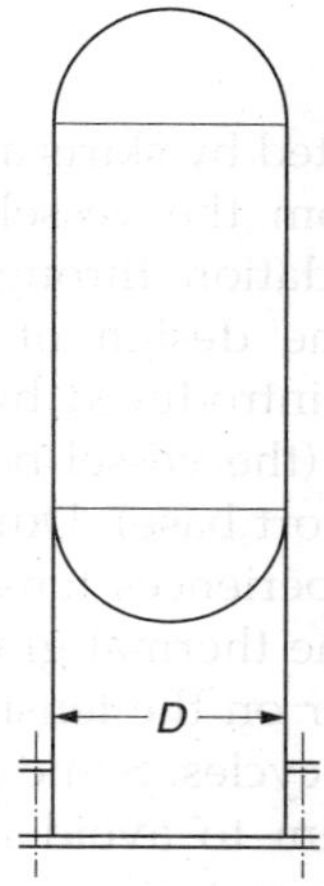

Figure 10.2 Support skirts.

$$P_r = \sum_{n=0,1,2}^{\infty} \sum_{m=0,1,3}^{\infty} P_{n,m} \cos n\phi \cos\left(\frac{m\pi x}{L}\right) \tag{10.1}$$

where L is the length of the shell, the term $P_{n,m}$ is the loading term. This representation is used for different forms of vessel loadings, where the direct and moment loadings are expressed as double Fourier series and introduced into the shell equations to obtain the values of stress resultants and displacements. In order to represent the patch load from the lug it is often necessary to have a large number of terms (typically about 200) in both the circumferential and axial directions. This approach has been used to draw up the curves presented in WRC 107.[2] When the attachment contact face is not rectangular, but maybe circular or elliptical, the design codes attempt to resolve the geometry into a rectangular patch. Mirza and Gupgupoglu[3] have studied stresses and displacements in circular cylindrical shells having square and rectangular lugs separated 90° apart along the circumferential direction. They have utilized a finite-element technique using 17-node doubly curved shell elements. For values of $C_1 = C_2 = 0.1\,D$, and $D/t \leq 40$ (where D is the mean diameter of the shell, and t the thickness) there seems to be a good agreement of results between References 2, 3, and 4. However, for smaller values of C_1 and C_2 (typically less than 0.05 D), large variations occur between the finite-element results[3] and closed-form predictions of References 2 and 4. However in spite of the variations in predicting the magnitude of the maximum stresses the methods agree on the direction of the maximum stress.[3]

The maximum stress is located at the upper end of the lug-vessel interface and occurs at the outside surface of the shell.

10.3 Support skirts

Most vertical vessels are supported by skirts as shown in Figure 10.2. These supports transfer the loads from the vessel by shear action. They also transfer the loads to the foundation through anchor bolts and bearing plates. A major problem in the design of support skirts involves the consideration of thermal stress introduced by the thermal gradient of the skirt in contact with the vessel (the vessel being at a considerably higher temperature than the cold support base). During heating up of the vessel, the outside of skirt juncture experiences tension, the magnitude of which depends upon the severity of the thermal gradient along the length of the skirt. Fatigue cracks may appear on the tensile surface of the weld due to alternating heating and cooling cycles. Some of the features that enter into the design of such supports aim to avoid attachments with high stress concentrations, to avoid partial penetration welds, and to employ generous fillet radii. In order to reduce the axial thermal gradient, suitable insulation must be adopted.

Skirt construction permits radial growth of pressure vessel due to pressure and temperature through the bending of skirt acting like a beam on an elastic foundation. The choice of proper height of the skirt support ensures that bending takes place safely. Finite-element methods can be effectively used to determine the stresses and deflections due to imposed pressure and temperature distribution.

In order to design a skirt support for mechanical loads alone, consider the vessel deadweight, W, and the bending moment, M, produced by seismic, wind and other mechanical loads. The stress in the skirt is then a combination of axial and bending stresses and is given by

$$\sigma = \frac{-W}{A} \pm \frac{M}{Z} \tag{10.2}$$

where A is the cross-sectional area and approximately equals πDt, with D the diameter and t the thickness. Z is the section modulus and for the circular cross-section is approximately equal to $0.25\pi D^2 t$. Eq. (10.2) then becomes

$$\sigma = \frac{-W}{\pi D t} \pm \frac{4M}{\pi D^2 t} \tag{10.3}$$

Once the thickness of the skirt, t, is determined, the next task is to design the anchor bolts. If the total number of anchor bolts are N, then the load on a single bolt, P, from the consideration of bending about a neutral axis is given by

$$P = \frac{-W}{N} \pm \frac{4M}{ND} \tag{10.4}$$

10.3.1 Example problem

Design a skirt support for a pressure vessel with a total vertical load of 720 kN, and an overturning moment of 2050 kNm. The bolt circle diameter of the support may be assumed to be 4.5 m. Assume a thickness of 10 mm for the support skirt and the mean diameter of the support as 4.25 m.

10.3.2 Solution

The support is to be designed such that it does not buckle under the compressive load. With the given dimensions, namely $D = 4.25$ m, $t = 10$ mm, $W = 720$ kN, and $M = 2050$ kNm we have

$$\sigma = \frac{-720 \times 10^3}{\pi(4250)(10)} - \frac{4 \times 2050 \times 10^3 \times 10^3}{\pi(4250)^2(10)} = -19.84 \text{ MPa} \tag{10.5}$$

The procedure outlined in Chapter 5 may be used to demonstrate that the skirt does not buckle for the compressive stress of 19.84 MPa, calculated in Eq. (10.5).

The anchor bolts are designed as follows. Let $N = 12$ bolts. The load per bolt is calculated using Eq. (10.4) as

$$P = \frac{720 \times 10^3}{12} + \frac{4 \times 2050 \times 10^6}{12 \times 4250} = 166.8 \text{ kN} \tag{10.6}$$

For class 4.6 metric steel bolts, the proof strength is 225 MPa.[5] The load determined in Eq. (10.6) when divided by this proof strength gives the required bolt area as 741 mm^2.

For ISO metric standard screw threads with a major diameter of 36 mm the stress area is 816.72 mm^2, which is closest to the required bolt stress area.[5] Therefore 12 bolts of 36 mm diameter can be used as the anchor bolts.

10.4 Saddle supports

Horizontal pressure vessels are usually supported on two symmetrically spaced saddle supports (Figure 10.3). Zick[6] presented a semiempirical analysis approach that has traditionally formed the basis of design of saddle supports for horizontal pressure vessels in a number of pressure vessel design codes. In this approach the vessel is assumed to behave as a simply supported beam, and the cross-section is assumed to remain circular under load. The simplified model as represented in Zick's analysis is an overhanging beam subjected to a uniform load due to the weight of the vessel and its contents (Figure 10.4). If the longitudinal bending moment, M_1, at the midspan of the vessel is equal to the longitudinal bending moment, M_2, at the saddle, then an optimum location of the saddle support can be obtained. The bending moments M_1 and M_2 are determined to be

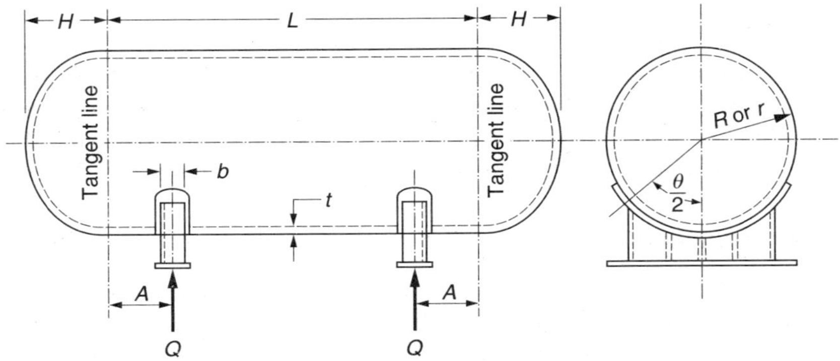

Figure 10.3 Saddle supports.

$$M_1 = q\left[\frac{(L-2A)^2}{8} - \frac{2}{3}HA - \frac{A^2}{2} + \frac{H^2}{4}\right] - M_0 \qquad (10.7)$$

and

$$M_1 = q\left[\frac{2}{3}HA + \frac{A^2}{2} + \frac{H^2}{4}\right] - M_0$$

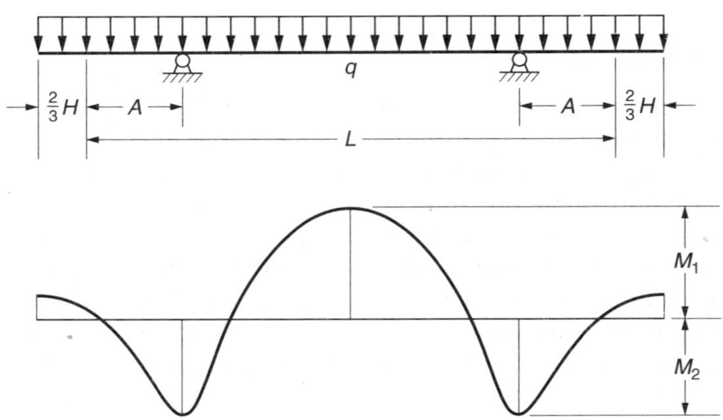

Figure 10.4 Simplified beam model for saddle supports.

Setting $M_1 = M_2$, from Eqs. (10.7) and (10.8) we obtain

$$\frac{(L-2A)^2}{8} - \frac{2}{3}HA - \frac{A^2}{2} = \frac{2}{3}HA + \frac{A^2}{2} \tag{10.9}$$

Furthermore, if we assume hemispherical heads for which $R = H$, and for the particular case of $L/R = L/H = 30$, we obtain the following relationship between A and L:

$$A^2 - 1.08LA - \frac{L^2}{4} = 0 \tag{10.10}$$

This gives:

$$\frac{A}{L} = 0.195 \tag{10.11}$$

Widera *et al.*[7] rightly note that the above optimum location of the saddle results from the consideration of the vessel as a beam, and not as a shell. In order to capture the shell behavior, they analyzed six vessel models using the finite-element method. These models correspond to A/L values of 0.05, 0.10, 0.15, 0.20, 0.25, and 0.30.[7] They concluded that for $A/L = 0.25$, the minimum values of stress intensities occur at the midspan, top line and bottom line of the vessel, as well as the local stresses at the saddle-vessel interface regions. Therefore they suggest that the optimum location of the saddle support is $A/L = 0.25$. Their other conclusions are that the displacements at the saddle and the midspan are in good agreement with the results obtained in Zick's analysis,[6] as far as the distribution and location of the maximum values are concerned; although the maximum values Widera *et al.* obtained are much smaller. Also, in the area of the saddle, high concentrated stresses and large stress gradients exist. Away from the region, the stresses decay very rapidly. In the region of the saddle-vessel interface, the stress distribution is not uniform, and the high localized stress represents a combination of membrane and bending stresses, and should be considered as primary plus secondary stresses.

It should be noted that the Zick's solution[6] is limited to the consideration of the dead weight of the horizontal vessel and contents for a rather specific saddle design configuration. Therefore in a general situation involving specific geometry or loading conditions, recourse must be made to finite-element methods. Horizontal pressure vessels are generally fixed at one saddle and the other is allowed to move to accommodate thermal expansion. The entire seismic load is therefore resisted by the fixed saddle, which can result in very high local bending stress on the shell due to the overturning moment on the saddle. A possible design modification is to add a series of gusset plates to the saddle, to reduce the overturning moment on the shell thereby reducing the stresses.

References

1. Bijlaard, P.P., Stresses from local loadings in cylindrical vessels, *Trans. Am. Soc. Mech. Eng*, 77, 805–818, 1995.
2. Wichman, K.R., Hopper, A.G., and Mershon, J.L., Local stresses in spherical and cylindrical shells due to external loadings, *Weld. Res. Counc. Bull. No. WRC-107*, 1965.
3. Mirza, S., and Gupgupoglu, K., Stress analysis of pressure vessels with uniformly spaced lugs, *J. Pressure Vessel Technol.*, 110, 70–75, 1988.
4. Rodabaugh, E.C., Dodge, W.G., and Moore, S.E., Stress indices at lug supports on piping systems, *Weld. Res. Counc. Bull. No. WRC-198*, 1974.
5. Norton, R.L., *Machine Design: An Integrated Approach*, 2nd ed., Prentice Hall, 2000.
6. Zick, L.P., Stress in large horizontal cylindrical pressure vessels on two saddle supports, *Weld. J. Res. Suppl.*, 435–445, 1951.
7. Widera, G.E.O., Sang, Z.F., and Natarajan, R., On the design of horizontal pressure vessels, *J. Pressure Vessel Technol.*, 110, 393–401, 1988.

chapter eleven

Simplified inelastic methods in pressure vessel design

Contents

11.1 Introduction

The ASME Code[1] has specific rules for the purpose of fatigue evaluation using elastic analyses. It is well known that many pressure vessels experience thermal transients, which lead to the fatigue cracks. The number of stress cycles applied during the specified life of these components seldom exceeds 100,000 and is frequently of the order of a few thousand. In this case, plastic behavior may be allowed during strain cycling, and accordingly the phenomenon is called low cycle fatigue. The local plastic strain controls fatigue crack initiation. Specifically the code design procedures are based on crack initiation, elastic stress analysis, and *S–N* curves for smooth specimen tests. The use of an elastic stress analysis to address the local strain range for low cycle fatigue analysis where significant plastic deformation takes place, warrants careful consideration. It seems a comprehensive inelastic analysis is the only way to address this issue. There are however, a number of difficulties associated with a detailed inelastic analysis. First, inelastic calculations are quite a bit more expensive than elastic ones, thus necessitating avoiding inelastic computations in design. More importantly however, the issues of constitutive modeling place a big hurdle in the way of comprehensive inelastic analysis. The analysis must account for cyclic hardening and softening along with

Bauschinger's effect. The ASME Code specifies some correction factors to be used in conjunction with elastic stress analysis that makes the evaluation plausible and in many cases conservative.

There are two specific sections of the ASME Code that employ these correction factors. Both sections attempt to address the differences between the values by elastic computation and the real value, but have two distinct origins. The first one relates to the differences in volume between elastic and inelastic behaviors. NB 3227.6 of the ASME Code suggests using a modified Poisson's ratio, ν, for elastic stress analysis as follows:[1]

$$\upsilon = 0.5 - 0.2S_{\mathrm{y}}/S_{\mathrm{a}} \tag{11.1}$$

Here S_{y} is the yield strength of the material at the mean value of the temperature of the cycle and S_{a} is the value obtained from the applicable design fatigue curve for the specified number of cycles of the condition being considered. The use of Eq. (11.1) is recommended for local thermal stresses only. Such an approach with a modified expression for Poisson's ratio involving the secant modulus, E_{s} has been employed in low cycle fatigue evaluation by Gonyea[2] and Moulin and Roche[3]. In their assessments the Poisson's ratio expression is represented as

$$\upsilon = 0.5 - (0.5 - \upsilon_{\mathrm{e}})E_{\mathrm{s}}/E \tag{11.2}$$

where E_{s} is the secant modulus obtained from the cyclic stress strain curve and E is the Young's modulus. ν_{e} represents the elastic Poisson's ratio whose value has been assumed to be 0.3 in Eq. (11.1).

Both Eqs. (11.1) and (11.2) account for the effect of transverse strain on plastic strain intensity factor characterized by the modified Poisson's ratio, ν. In Eq. (11.1), this is accounted for by the ratio $S_{\mathrm{y}}/S_{\mathrm{a}}$, whereas in Eq. (11.2) the ratio E_{s}/E serves the same purpose as will be shown later. The modified Poisson's ratio in each case is intended to account for the different transverse contraction in the elastic–plastic condition as compared to the assumed elastic condition. Therefore this effect is primarily associated with the differences in variation in volume without any consideration given to the nonlinear stress–strain relationship in plasticity. Instead the approaches are based on an equation analogous to Hooke's law as obtained by Nadai.[4] Gonyea[2] uses expression (rule) due to Neuber[5] to estimate the strain concentration effects through a correction factor, K_{ν} for various notches (characterized by the elastic stress concentration factor, K_{T}). Moulin and Roche[3] obtain the same factor for a biaxial situation involving thermal shock problem and present a design curve for K_{ν} for alloy steels as a function of equivalent strain range. Similar results were obtained by Houtman for thermal shock in plates and cylinders and for cylinders fixed to a wall,[6,7] which were discussed by Nickell.[8] The problem of Poisson's effect in plasticity has been discussed in detail by Severud.[9] Hubel has obtained a K_ν factor for a cylinder subject to a radial temperature gradient.[10]

The second correction factor in the ASME Code focuses on the nonlinear stress–strain relationship. This implies that the actual strain variation is generally greater than the value computed based on the assumption of elastic material behavior. NB 3228.3 of the Code is based on such consideration and introduces the K_e factor when the range of primary and secondary stress exceeds three times the S_m limit (typically twice the yield strength). This factor accounts for redistribution of local strains due to plastic flow near the stress concentration for stress ranges near to three times the S_m limit, and plastic redistribution of nominal strains for stress ranges exceeding this limit. Since the low cycle fatigue crack initiation is judged to be local plastic strain controlled, a correction factor must be applied to the elastically calculated stress to account for local and nominal strain distributions when operating at nominal stress ranges near or above the three times S_m limit. NB 3228.3 of the Code accounts for plastic strain redistribution by introduction of the K_e factor which is defined as

$$K_e = 1.0(S_n < 3S_m) \tag{11.3a}$$

$$K_e = (1-n)/[n(m-1)][S_n/3S_m - 1](3S_m < S_n < 3mS_m) \tag{11.3b}$$

$$K_e = 1.0/n(S_n > 3mS_m) \tag{11.3c}$$

where n is the strain hardening exponent and m is a fitting parameter. Both are considered material properties. Note that the K_e factor depends on the nominal stress range, S_n and three material parameters: S_m, m, and n. There are no parameters related to the geometry of the component in question.

The evolution of the K_e factor has been based on simple analyses and limited test data. Krempl studied the low cycle fatigue behavior of notched cylinders and plates of three pipe materials.[11] Tagart formulated design rules for the Nuclear Piping Code B31.7 based partly on these tests,[12] which led to Eq. (11.3b). Langer[13] performed elastic–plastic analyses of a beam in bending and a tapered bar in tension which led to Eq. (11.3c).

Petrequin, Roche and Tortel derived a plastic strain magnification factor (similar to the ASME Code K_e factor) for notches using Neuber's rule and cyclic stress–strain curves.[14] This factor is calculated as a function of elastic strain range.

Safety margins of Eqs. (11.3a) to (11.3c) were evaluated by Gerber[15] and by Iida *et al.*[16] Gerber performed low cycle fatigue tests on type 304 stainless steel and A516 steel and showed that the ASME Code K_e factors were always larger than the values obtained experimentally.[15] However the finite-element analysis performed by Gerber on test conditions showed that the K_e values tend to be larger than the ASME Code values, when the nominal stress range is close to $3S_m$ limit. Iida *et al.*[16] experimentally demonstrated that the K_e values of the ASME Code were conservative when the nominal stress ranges were significantly larger than $3S_m$, but could be nonconservative when the nominal stress ranges were close to $3S_m$. This aspect is apparently recognized in the MITI Code[17] where a correction has been introduced at $S_n = 3S_m$. Chattopadhyay[18] employed the local strain approach to assess the ASME and MITI Codes for this feature and suggested improvement needed in the ASME Code for low cycle fatigue evaluations.

11.2 Elastic analysis incorporating modified Poisson's ratio

NB 3227.6 of the ASME Code Section III,[1] prescribes variable Poisson's ratio for calculating local thermal stresses through Eq. (11.1). An objective of using a variable Poisson's ratio is to help overcome the underprediction of the equivalent strain range obtained using elastic methods where a Poisson's ratio of 0.3 is used. In order to perform a fatigue analysis for the case of a thermal shock, it is necessary to determine the strain along the surface as well as perpendicular to the surface. This component arises from the Poisson effect, namely the shrinkage in one direction when elongation is caused in two other directions. In other words, the effect represents the volume variation. These variations differ according to whether the strain is elastic or not. Moulin and Roche present the K_ν factors as a function of the equivalent strain range using the secant modulus and the effective Poisson's ratio as given in Eq. (11.2).[3]

In the following, we present an analysis performed using the NB 3227.6 expression for the Poisson's ratio as given in Eq. (11.1). For this analysis we assume a uniform flat plate subjected to a temperature gradient $\triangle T$ throughout while being held at edges. The elastically computed strain range, $\triangle \varepsilon$, for this case is given by:

$$\Delta\varepsilon = \alpha\Delta T/(1-\upsilon) \tag{11.4}$$

In Eq. (11.1), replacing S_y with $3/2S_m$ and S_a by $S_n/2$, we get

$$\upsilon = 0.5 - 0.2(3S_m/S_n) \tag{11.5}$$

If we indicate the inelastically computed strain range by $\triangle \varepsilon^*$ and the elastically computed strain range by $\triangle \varepsilon$, then it follows that

$$\Delta\varepsilon^* = \Delta\varepsilon(1-0.3)/[1-\{0.5-0.2(3S_m/S_n)\}] \tag{11.6}$$

Thus the factor K_ν is given by

$$K_\upsilon = \Delta\varepsilon^*/\Delta\varepsilon = 7/[5+2(3S_m/S_n)])] \tag{11.7}$$

In Figure 11.1, the factor K_ν is presented as a function S_n/S_m of using Eq. (11.6). A similar figure has been presented by Severud[9] using a similar expression for the Poisson's ratio, ν, in Eq. (11.5) and employing deformation theory of plasticity. Severud[9] also compares his results with those obtained by Houtman[6,7] using the finite-element method.

The results of Moulin and Roche[3] are also shown in Figure 11.1 by setting

$$\Delta\varepsilon/\Delta\varepsilon_y = S_n/S_m \tag{11.8}$$

where ε_y is the yield strain taken as 0.002. The results obtained by Houtman[6] and by Severud[9] are also indicated in Figure 11.1. The plastic strain concentration factors obtained by Hubel[10] are also indicated in the same figure. Hubel's results are based on a hardening coefficient of zero, or elastic–perfectly plastic material behavior, and the value of K_ν is obtained as

$$K_\upsilon = 1 + (1 - 2\upsilon_e)(1 - 3S_m/S_n) \tag{11.9}$$

The estimates of K_ν provided by Eq. (11.7) (11.9) due to Hubel[10] are in close agreement and both reach asymptotically the same value of 1.4 for large vales of S_n/S_m. The results obtained by various investigators indicate that procedure specified in ASME Code Subsection NB 3227.6 provide reasonable estimates for the plastic strain intensification for local thermal stresses. The simplified approach leading to Eq. (11.7) provides a lower bound estimate for K_ν as seen in Figure 11.1. The results due to Moulin and Roche[3] show the K_ν values, greater than unity for $S_n/3S_m$ values less than one, primarily because their estimates are strain-based, and indicates plastic deformation occurring at or before yield strength.

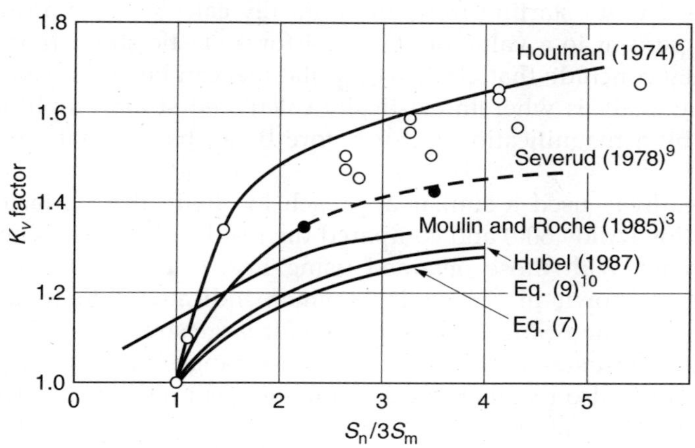

Figure 11.1 K_ν factor vs. stress range.

11.3 Elastic analysis to address plastic strain intensification

ASME Code Subsection NB 3228.3 prescribes simplified elastic–plastic analysis that applies fatigue enhancement factors to the elastically computed strain range, when the primary-plus-secondary stress range, S_n, exceeds the $3S_m$ limit. (S_m approximately equals two-thirds of the material yield strength or one third of the ultimate strength, whichever is higher.) For primary-plus-secondary stress ranges below the $3S_m$ limit, shakedown of the structure to elastic behavior is assumed and the knowledge of the elastic stress concentration factor, K_t, and information about the other loading parameters, is judged adequate for low cycle fatigue design. When the $3S_m$ limit is exceeded, the strict shakedown limit is exhausted leading to the presence of cycles of alternating yielding in zones of stress concentration. There is redistribution of local strains due to plastic flow at those zones. Therefore it is necessary to perform a low cycle fatigue analysis, because the low cycle fatigue crack initiation is local plastic strain controlled. A correction factor must be applied to the elastically calculated stresses to account for local and nominal strain redistributions when designing for nominal stress ranges near or above the $3S_m$ limit. The K_e factor as indicated in Eqs. (11.3a) to (11.3c) essentially gives the elastic–plastic strain ranges from the strain ranges calculated by purely elastic means. Alternatively, Eqs. (11.3a) to (11.3c) provide for decay in fatigue lives when cycling in the inelastic range.

Petrequin, Roche and Tortel[14] derived a plastic strain intensity factor for notches by introducing a cyclic hardening law based on two material constants and use the approach due to Neuber,[5] which assumes the invariance of the product of stress and strain for both the linear elastic behavior and the true one. They have obtained the K_e: factors for 316L stainless steels as a function of elastically calculated strain range. The factors exceed unity starting from an elastically calculated strain range as little as 0.1 percent to a value of about 1.4 for a elastic strain range of 0.3 percent. They conclude that elastic computations can be employed for low cycle fatigue analysis wherein elastically calculated strain amplitudes are multiplied by a magnification factor before being used in fatigue design curves.

Chattopadhyay used a similar approach to obtain the K_e: factor for a structural alloy (Alloy 600) and compared the results obtained using a local strain approach with those obtained using Eqs. (11.3a) to (11.3c).[19] The comparison is shown in Figure 11.2, which indicates a high degree of conservatism for the ASME Code approach for large stresses and indicating mild nonconservatism of the code for stress ranges close to $3S_m$. Such nonconservatism also results from the finite-element results of Gerber[15] for specimens at the locations of fillet and shoulder. This is consistent with Neuber's correction,[5] which indicates that when local yielding occurs, the

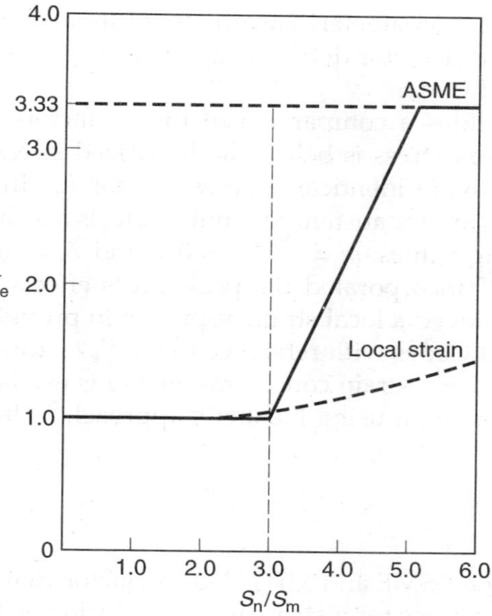

Figure 11.2 Comparison of K_e factors: ASME vs. local strain.

strain correction needs to be applied to the elastically calculated strains (and hence stresses). However Gerber did not observe nonconservatisms of the ASME Code equations experimentally.[15]

The experimental results of Iida *et al.*[16] indicate that the K_e: factors are conservative when the nominal stress range is significantly larger than the $3S_m$ limit, but could be nonconservative when the nominal stress range is close to $3S_m$.

When the primary-plus-secondary stress range is close to $3S_m$, there will be small plastic zones well contained by the remainder of the structure, which is assumed to stay elastic. The ASME Code ignores the contribution of the localized plastic zone as long as the range of stress intensity is less or equal to $3S_m$. Note that the stress intensity ranges are linearized values through the thickness of the cross-section. This means that even when the stress intensity range (linearized through the section thickness) is equal to $3S_m$, the surface peak stress intensity range, S_p can exceed $3S_m$, thus enhancing the potential to initiate surface cracks.

The MITI Code[17] apparently recognizes this feature and provides a strain enhancement factor for stress ranges, S_n in the neighborhood of $3S_m$, where values of K_e exceed unity and progressively increase with increasing ratios of S_p/S_n, where S_p is the peak. Equation (11.3b), valid for primary-plus-secondary stress intensity ranges between $3S_m$ and $3mS_m$, is modified in the MITI Code[17] as

$$K_e = 1 + A_o(S_n/3S_m) - (S_n/S_p) \tag{11.10}$$

where A_o is a material parameter. The greater of the two expressions given in Eqs. (9.3b) (9.10) is used for determining K_e factor in the region $3S_m \leq S_n \leq 3mS_m$ in the MITI Code.

Figure 11.3 provides a comparison of the K_e factors used in the two codes. When the peak stress is below the linearized stress, the ASME and the MITI Codes provide identical expressions for K_e. In Figure 11.3, the material parameters are for austenitic stainless steels and high nickel alloys, having the following values: $m = 1.7$, $n = 0.3$, and $A_o = 0.7$.

Chattopadhyay[18] incorporated the peak stress effects addressing them as notches and employed a local strain approach to provide an appraisal of the ASME and MITI Codes with reference to the K_e factor. In that work, the ratio S_p/S_n is treated as a strain concentration and is used as a square of the elastic stress concentration using Neuber's approach.[5] Thus

$$K_t^2 = S_p/S_n \tag{11.11}$$

In Figure 11.4, the ASME and MITI Code K_e factors have been compared with the computed values for a structural alloy (Alloy 600) using the cyclic stress–strain curve and using $S_p/S_n = 2.0$ and 5.0. The computed values of the K_e factors obtained by Chattopadhyay[18] appear to follow the same trend as the ones specified in the MITI Code for the material investigated.[17]

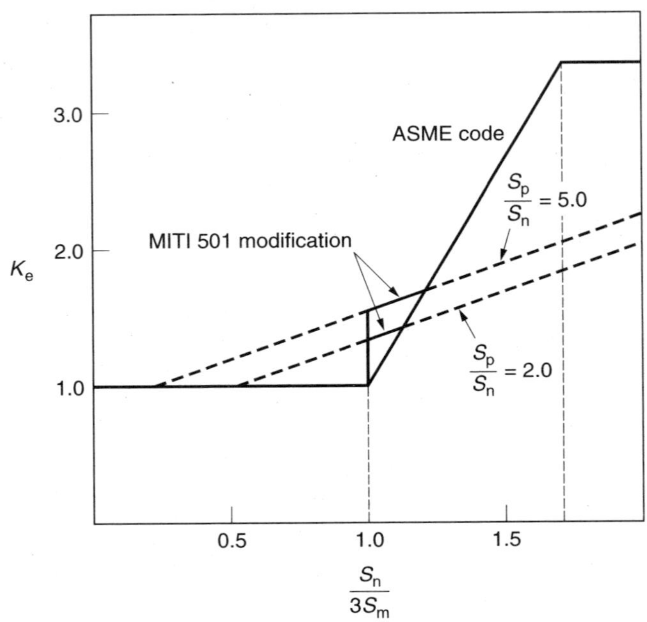

Figure 11.3 Comparison of K_e factors: ASME vs. MITI.

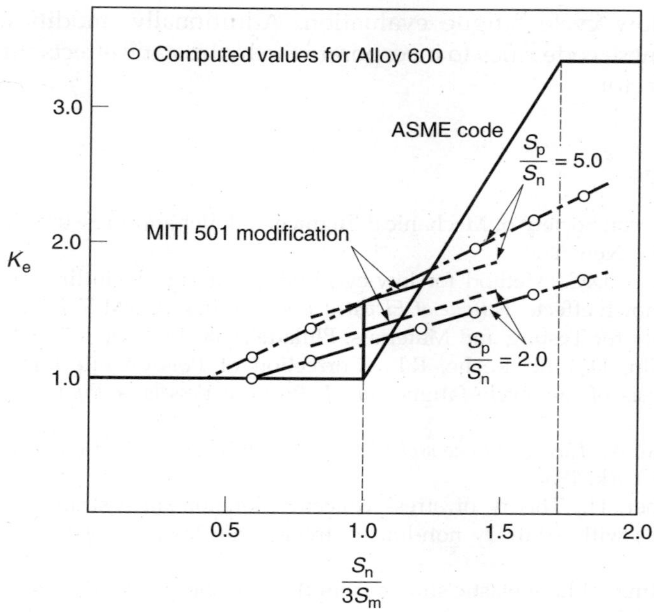

Figure 11.4 Comparison of K_e factors: computed vs. design codes.

11.4 Conclusion

In summary, the simplified inelastic analysis rules as indicated in subsections NB 3327.6 and NB 3228.3 of the ASME Boiler and Pressure Vessel Code Section III have been critically appraised. The first rule is shown to be equivalent to a correction factor, K_ν to be applied to local thermal stresses, and is based on an analysis involving a modified Poisson's ratio. For a simplified situation of thermal stress in a plate with a through the thickness temperature gradient (perfect biaxiality) the solution using NB 3227.6 are comparable to the existing solutions in the literature. However, the solutions obtained using finite-element methods and a different form of Poisson's ratio than that specified in NB 3227.6 (Eq. (11.1)) typically yield higher values of K_ν.

The second rule (NB 3228.3) on the K_e factors is judged to be fairly conservative for large stress ranges; however, it tends to be on the nonconservative side for values of stress intensity ranges equal or slightly greater than the $3S_m$ limit. Modification along the same lines as the MITI Code is suggested. Also, additional test data are needed in this area for a number of structural materials to properly validate the code equations.

It seems worthwhile to use both the factors K_e and K_ν in a typical application by multiplying the product of these two factors with the elastically calculated stresses before entering stresses in a design fatigue curve for low cycle fatigue evaluation. Additionally, modifications are needed to these code rules to incorporate the hydrostatic effects through the triaxiality factor.

References

1. American Society of Mechanical Engineers, Boiler and Pressure Vessel Code, ASME, New York.
2. Gonyea, D.C., Method for low-cycle fatigue design including biaxial stress and notch effects, *Fatigue at Elevated Temperatures,* ASTM STP 520, American Society for Testing and Materials, Philadelphia, 1973, pp. 673–687.
3. Moulin, D., and Roche, R.L., Correction of Poisson effect in the elastic analysis of low-cycle fatigue, *Int. J. Pressure Vessels & Piping,* 19, 213–233, 1985.
4. Nadai, A., *Theory of Flow and Fracture of Solids,* Vol. 1, 2nd ed., McGraw-Hill, New York, 1950.
5. Neuber, H., Theory of stress concentration for shear-strained prismatical bodies with arbitrary non-linear stress–strain law, *J. Appl. Mech.,* 544–550, 1961.
6. Houtman, J.L., Inelastic strains from thermal shock, *Machine Design,* 46, 190–194, 1974.
7. Houtman, J.L., Elastic–plastic strains from thermal shock at a discontinuity – a simplified approach, *Simplified Methods in Pressure Vessel Analysis,* ASME Publication PVP-PB-029, Barsoum, R.S., Ed., pp. 1–12, 1978.
8. Nickell, R.E., A Survey of Simplified Inelastic Methods, *Weld. Res. Counc. Bull.* 253, 1979.
9. Severud, L.K., Background to the elastic creep-fatigue rules of the ASME B&PV Code case 1592, *Nucl. Eng. Design,* 45, 449–455, 1978.
10. Hubel, H., Plastic strain concentration in a cylindrical shell subjected to an axial or radial temperature gradient, *J. Pressure Vessel Technol.,* 109, 184–187, 1987.
11. Krempl, E., Low cycle fatigue strength reduction in notched flat plates, *ASME Paper 67-MET-13,* ASME, 1967.
12. Tagart, S.W., Plastic fatigue analysis of pressure components, *ASME Paper 68-PVP-3,* ASME, 1968.
13. Langer, B.F., Design stress basis for pressure vessels, *Exp. Mech.,* 1–12, 1971.
14. Petrequin, P., Roche, R., and Tortel, J., Life prediction in low cycle fatigue using elastic analysis, *Proceedings of an ASME Conference on Advances in Life Prediction Methods,* Albany, New York, 1983, pp. 151–156.
15. Gerber, T.L., *Effect of Constraint Mode and Loading Mode on Low Cycle Fatigue Crack Initiation – Comparison with Code Design Rules, GEAP-20662,* UC-78, AEC Research and Development Report, General Electric Company, 1974.
16. Iida, K., Kitagawa, M., Tamura, K., Matsushita, A., Fukagawa, M., and Saiga, Y., Safety margin of the simplified elasto plastic fatigue analysis method of ASME Band PV Code Section III, *Proceedings of the Third International Congress on Pressure Vessel Technology, Institution of Mechanical Engineers,* London, 1980. Paper C35/80.
17. Ministry of International Trades and Industry, *Structural Design Rules for Nuclear Power Plant Components,* Notification 501, MITI, Japan, 1980.

18. Chattopadhyay, S., Multi axial low cycle fatigue evaluations of pressure vessel components, in *Advances in Multiaxial Fatigue*, McDowell, D.L., and Ellis, R., Eds., STP 1191, American Society of Testing and Materials, Philadelphia, 1993, pp. 412–428.
19. Chattopadhyay, S., A comparison of fatigue damage estimates using the ASME Code and local strain approaches, *ASME Paper 84-W-AfDE-7*, ASME, 1984.

chapter twelve

Case studies

Contents

12.1 Introduction

In the preceding chapters, the design procedures for various components of pressure vessels were outlined. In this chapter, some typical design examples will be presented. Some of the designs are based on the use of formulas (closed-form mathematical expressions) derived in the earlier chapters for shells and nozzles. The others involve finite-element analyses and the use of the stress results to demonstrate adequacy for primary and secondary stresses. Two of the examples involve fatigue evaluation. The inelastic fatigue evaluation of steam generator nozzle has been carried out using the ASME Code and the so-called local strain approach commonly used in the aerospace and automotive industries. The specific design and analysis efforts include

1. Sizing of thickness for a pressure vessel
2. Reinforcement calculations for a nozzle

3. Fatigue evaluation for a nozzle for a reactor pressure vessel (elastic analysis)
4. Fatigue evaluation for a nozzle for a steam generator (inelastic analysis)
5. Structural evaluation of a reactor vessel support

12.2 Sizing of a pressure vessel

The equations for cylindrical shells were presented in Chapter 5. The following equations have been used for designing a cylindrical shell as shown in Figure 12.1.

12.2.1 Example problem

This is an example of design of a nuclear reactor pressure vessel. The vessel shell has an inside shell radius of 2 m with a thickness of 200 mm and a clad thickness of 4 mm. The vessel is required to contain an internal pressure of 17 MPa. The object is to determine whether there is enough vessel thickness to contain the internal pressure.

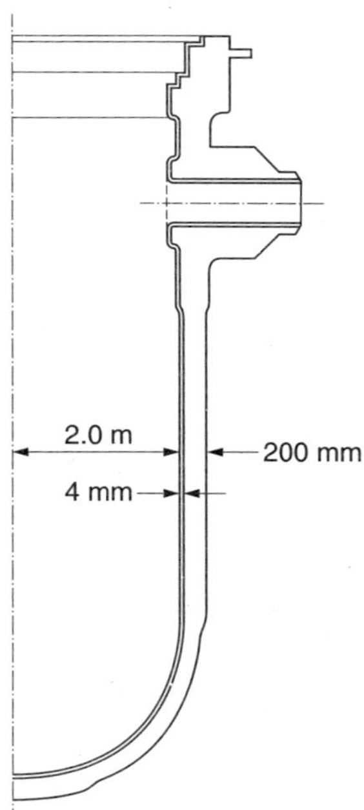

Figure 12.1 Vessel thickness sizing.

12.2.2 Solution

The material of the vessel is SA 533 Gr. B: Class 1 for which the allowable stress intensity (S_m.) from the ASME Boiler and Pressure Vessel Code, Section III is 184 MPa. We can take $P = 17$ MPa, $R = 2$ m $= 2000$ mm. From Eq. (5.15) with the joint efficiency E, taken as 100 percent or unity, and $S = S_m$ (this equation also appears in ASME Code Section III, Paragraph, NB 3324.1):

$$t_{\text{required}} = \frac{PR}{S_m - 0.5P} = \frac{17(200)}{184 - 0.5(17)} = 194 \text{ mm} \tag{12.1}$$

Since, $t_{\text{required}} < t_{\text{actual}} = 200$ mm, the thickness is adequate and the code criterion is met.

12.3 Nozzle reinforcement assessment

The subject of nozzles and openings and associated reinforcement requirements was discussed in Chapter 7. In this section, reinforcement calculations were presented for a vessel nozzle attachment. The reinforcement is provided by welding a reinforcing pad of width 250 mm and a thickness 150 mm as shown in Figure 12.2. There are two 80-mm-thick weld groups that also provide the reinforcement. Furthermore, the nozzle extends inside the shell accounting for further reinforcement. As shown in Figure 12.2, the shell thickness is 232 mm and the nozzle thickness is 76 mm. The inside radius of the nozzle is 375 mm to the base metal, and 368 mm to the cladding. The shell inside radius is 1.977 m to the cladding and 1.981 m to the base metal. An internal pressure of 17.2 MPa is applied to the shell and the nozzle. The material of the nozzle is SA 508, Cl–2 for which S_m = 184 MPa (from the ASME Boiler and Pressure Vessel Code, Section III). The required thickness of the shell and nozzle are calculated first.

12.3.1 Vessel and nozzle thickness calculations

The required vessel thickness (t_{vr}) is given by (using Eq. (5.15) with weld joint efficiency $E = 1$ or 100 percent and $S = S_m$):

$$t_{vr} = \frac{PR}{S_m - 0.5P} = \frac{17.2(1977)}{184 - 0.5(17.2)} = 194 \text{ mm} \tag{12.2}$$

The required nozzle thickness (t_{nr}) is given by (using Eq. (5.15) with weld joint efficiency $E = 1$ or 100 percent and $S = S_m$):

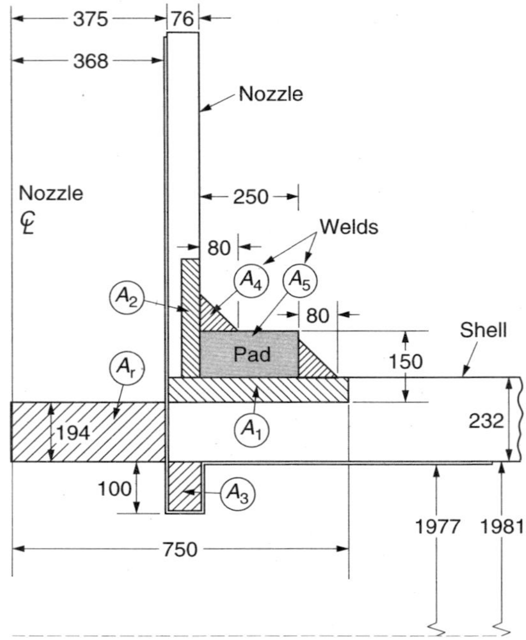

Figure 12.2 Nozzle reinforcement.

$$t_{nr} = \frac{PR}{S_m - 0.5P} = \frac{17.2(368)}{184 - 0.5(17.2)} = 36 \text{ mm} \tag{12.3}$$

12.3.2 *Reinforcement calculations*

To determine whether an opening is adequately reinforced, it is required to calculate the reinforcement areas available in the geometry. The appropriate equations are also summarized in Reference 1.

The required reinforcement area (A_R) for any plane through the center of the opening is given by the following equation (where d is the diameter of the opening (in this case the inside diameter of the nozzle) and t_{vr} is the required thickness of the vessel):

$$A_R = dt_{vr} = 2[375(194)] = 14550 \text{ mm}^2 \tag{12.4}$$

The limits of reinforcement parallel to the vessel surface measured on each side of the center line of the nozzle are the larger of d or $t_v + t_n + 0.5d$, with t_v and t_n being the thicknesses of the vessel and the nozzle respectively (ASME Code Section VIII Division 1 rule). In our case d = 750 mm, t_v =

232 mm, and $t_n = 76$ mm. Therefore the limit of reinforcement parallel to the vessel surface measured on each side of the center line of the nozzle is 750 mm (*d*). This is indicated in Figure 12.2.

The limit of reinforcement normal to the vessel measured outward from the surface of the vessel is the smaller of $2.5t_v$ or $2.5t_n + T_p$, where T_p is the thickness of the reinforcing pad.[1] In this case, $T_p = 150$ mm, $t_v = 232$ mm, and $t_n = 76$ mm. Therefore the limit of reinforcement normal to the vessel is given by $2.5t_n + T_p = 2.5(76) + 150 = 340$ mm. This forms the basis of available area in the nozzle. The area of reinforcement available includes:

The area of excess thickness in vessel, A_1
The area of excess thickness in nozzle wall, A_2
The area available in the nozzle projecting inward, A_3
The cross-sectional area of the welds, A_4
The cross sectional area of the reinforcing pad, A_5

The opening is adequately reinforced if:[1,2]

$$A_{total} = A_1 + A_2 + A_3 + A_4 + A_5 > A_R \tag{12.5}$$

The area available for reinforcement is calculated as follows:

Shell excess area, $A_1 = 2[(750 - 375)(232 - 194)] = 28500\ \text{mm}^2$

Nozzle excess area, $A_2 = 2[((2.5 \times 76) + 150)36] = 24480\ \text{mm}^2$

Nozzle extension (into the shell) area, $A_3 = 2[100 \times 76] = 15200\ \text{mm}^2$

Weld area, $A_4 = 2[0.5 \times (80^2 + 80^2)] = 12800\ \text{mm}^2$

Reinforcement pad area, $A_5 = 2[250 \times 150] = 75000\ \text{mm}^2$

$A_{total} = A_1 + A_2 + A_3 + A_4 + A_5 = 155980\ \text{mm}^2 > A_R = 145500\ \text{mm}^2$

This is acceptable from the standpoint of the ASME Section VIII, Division 1 rules.[1]

12.4 Fatigue evaluation using elastic analysis

The stress histogram for the structural component is shown in Figure 12.3. The stresses have been calculated using a finite-element structural analysis, and the peak stresses have been reported. All the stresses ranges are within the $3S_m$ limit of the material and elastic analysis forms the basis of the evaluation. A total of 18 transients (*a* through *r*) and an associated number of cycles are indicated. The maximum S_{max} and S_{min} of stresses for various transient combinations are indicated in Table 12.1. The alternating stress is

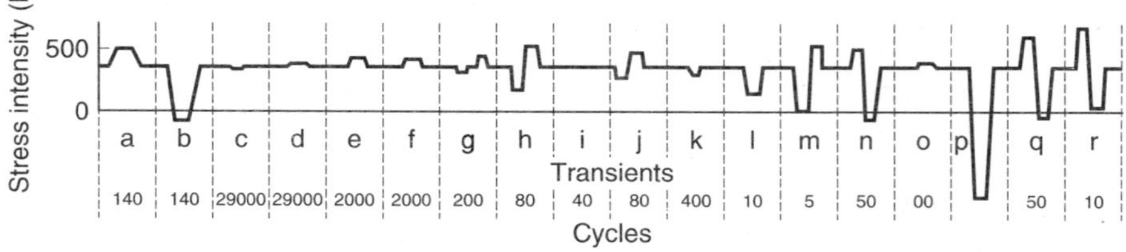

Figure 12.3 Calculated stress intensities for various transients.

Table 12.1 Fatigue Evaluation

Transients	S_{max}	S_{min}	S_{alt}	N	N	U
r–p	655.4	–659.6	657.5	750	1	0.0013
r–b	655.4	–82.5	369.0	3700	9	0.0024
q–b	595.9	–82.5	339.2	4600	50	0.0109
a–b	520.2	–82.5	301.4	6000	81	0.0135
h^+–n	509.9	–47.0	278.5	7900	50	0.0063
h^+–m	509.9	–4.0	257.0	11000	5	0.0005
h^+–l	509.9	134.9	187.5	32000	10	0.0003
h^+–h^-	509.9	194.4	157.8	65000	15	0.0002
j^+–j^-	444.3	265.5	89.4	∞	80	0.0000
	Total usage factor = 0.0354					

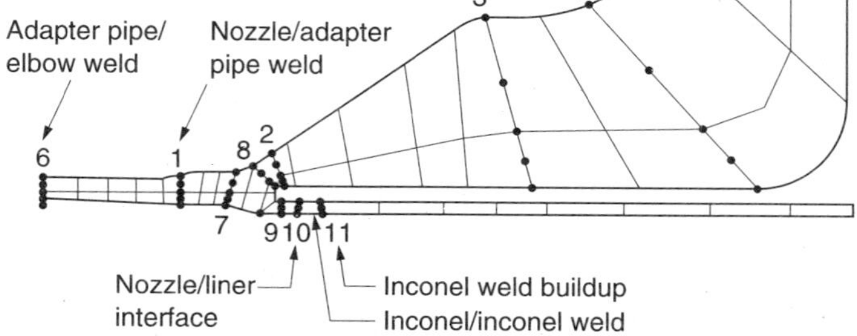

Figure 12.4 Finite-element model of a feedwater nozzle.

one half of the magnitude of the range, S_{alt}. Based on S–N curve in Figure 12.4 (dashed line), the allowable number of cycles, N, is obtained. The information on transient cycles provides the actual number of cycles, n. For each transient pair, the ratio n/N is calculated. The process involves giving in from the largest stress range progressively to smaller stress ranges, until the allowable number of cycles, N, goes to infinity, corresponding to the endurance limit.

A cumulative usage factor of 0.0354 is obtained. This value is well below unity demonstrating adequacy of nozzle for cyclic operation.

12.5 *Fatigue evaluation using the simplified inelastic analysis method*

The following design problem was studied to determine the effects of inelastic action on the fatigue evaluation. The detailed description of the procedure may be found in Reference 3. The finite-element model used for the design evaluation is shown in Figure 12.4. This shows a thermal liner welded to a nozzle. The loading conditions for the nozzle-to-liner weld are characterized by internal pressure, as well as temperature and flow excursions of the fluid entering the nozzle. The temperature distribution through the thickness of the weld was calculated using a finite-element model for a number of loading transients. The transient analyses involve specification of appropriate heat transfer coefficients at the fluid–solid interface. In order to provide peak stress conditions for fatigue evaluations, the times during the transient analysis have been identified by investigating the conditions when the shock and bending stresses are maximized. The thermal solutions at these instants of time have been used for subsequent finite-element elastic stress analyses. The calculated stresses at the outside surface of the analysis section 11 (Figure 12.4) associated with various loading conditions are listed in Table 12.2.

Based on the principal stresses reported in Table 12.2, the stress intensities and the effective stresses are computed. The stress intensity is the largest absolute magnitude of the difference in the principal stresses (σ_1, σ_2, σ_3) while the effective stress is given by

$$S_e = \frac{1}{\sqrt{2}}\left[(\sigma_1 - \sigma_2)^2 + (\sigma_2 - \sigma_3)^2 + (\sigma_3 - \sigma_1)^2\right]^{\frac{1}{2}} \tag{12.6}$$

Although the ASME Code procedure does not require the effective stresses for fatigue evaluation, these have been nonetheless calculated for comparison purposes. The stress intensities and the effective stresses are listed in Table 12.3 for all loading conditions. A sign has been applied to each of the stresses based on the inspection of the stress state associated with the particular loading conditions in Table 12.2. From Table 12.3, it can

Table 12.2 Loading Condition and Principal Stresses

Condition	Transient	Number of cycles	Principal stress, σ_1 (MPa)	Principal stress, σ_2 (MPa)	Principal stress, σ_3 (MPa)
1	Load	500	194	–126	–122
2	Normal	200	203	–105	–97
3	Upset 1	770	–190	–20	–32
4	Upset 2	20	249	–171	–163
5	Upset 3	30	364	–134	–123
6	Upset 4	30	265	–195	–185

Table 12.3 Stress Intensities and Effective Stresses

Condition	Stress intensity (MPa)	Effective stress (MPa)
1	319	318
2	308	304
3	–170	165
4	420	416
5	498	492
6	461	456

be noted that the stress intensities and the effective stresses are nearly the same for all loading conditions considered in this problem.

The material for the thermal liner as well as the liner-to-nozzle weld is Inconel 600, for which the modulus of elasticity, E, and the yield strength, S_y, are 207 GPa and 240 MPa, respectively. The design stress intensity S_m for the material is 2/3 S_y or 160 MPa. The design fatigue curve for the material is obtained from Reference 2 and is shown in Figure 12.5. This figure is based on an elastic modulus of 180 GPa. The alternating stresses

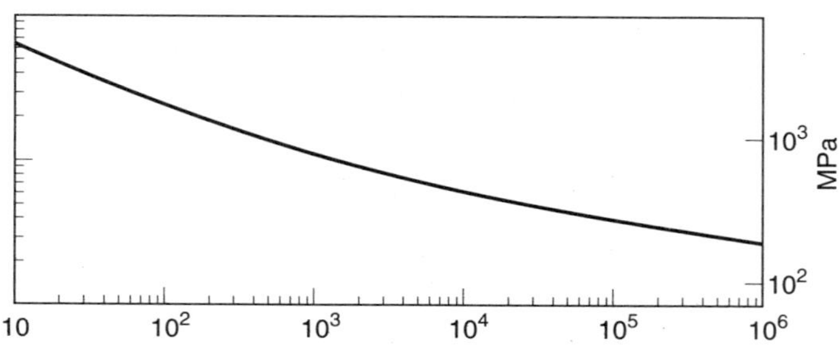

Note: $E = 26.0 \times 10^6$ psi (1.8×10^5 MPa)

Figure 12.5 Fatigue curve for nozzle liner material.

Table 12.4 Fatigue Evaluation, $K = 0.87$

Load combination	Stress intensity range (MPa)	Ratio (linearized to surface SI)	K_e factor (based on linearized SI)	Stress amplitude (MPa) = ½ (SI range)(KK_e)	No. of cycles (n)	Allowable no. of cycles (N)	Fatigue usage factor (n/N)
5–3	668	1.41	3.33	971	30	450	0.067
6–3	632	1.35	3.33	918	30	534	0.056
4–3	591	1.36	3.20	825	20	747	0.027
1–3	490	1.37	2.32	501	500	4345	0.115
2–3	479	1.40	2.31	483	190	4994	0.038

Cumulative fatigue usage factor $\Sigma\ (n/N) = 0.303$.

have to be multiplied by a factor of $K = 180/207$ or 0.87 to perform the fatigue evaluation.

Table 12.4 shows the fatigue evaluation results using the ASME Code procedure. All of the stress ranges exceed the $3S_m$ limit (twice the yield strength of the material) necessitating the use of a simplified elastic–plastic analysis, and the associated computation of the K_e factor (using Eqs. (11.3a), (11.3b) or (11.3c) as appropriate). A cumulative fatigue usage factor of 0.303 is obtained from this evaluation.

12.6 Structural evaluation of a reactor vessel support

The reactor vessel support configuration is indicated in Figure 12.6 and the detailed description may be found in Reference 4. This has been designed to the requirements of the "Component Support" subsection of the ASME Boiler and Pressure Vessel Code, Section III. This consists of a ring structure with a box cross-section supporting the reactor vessel to the reactor cavity support ledge (Figure 12.6). Long hold-down bolts, which pass through holes in the reactor vessel flange, the vessel support and the ledge, secure the reactor vessel to the ledge and clamp the support between them. The box-ring support structure is comprised of top and bottom low alloy-steel

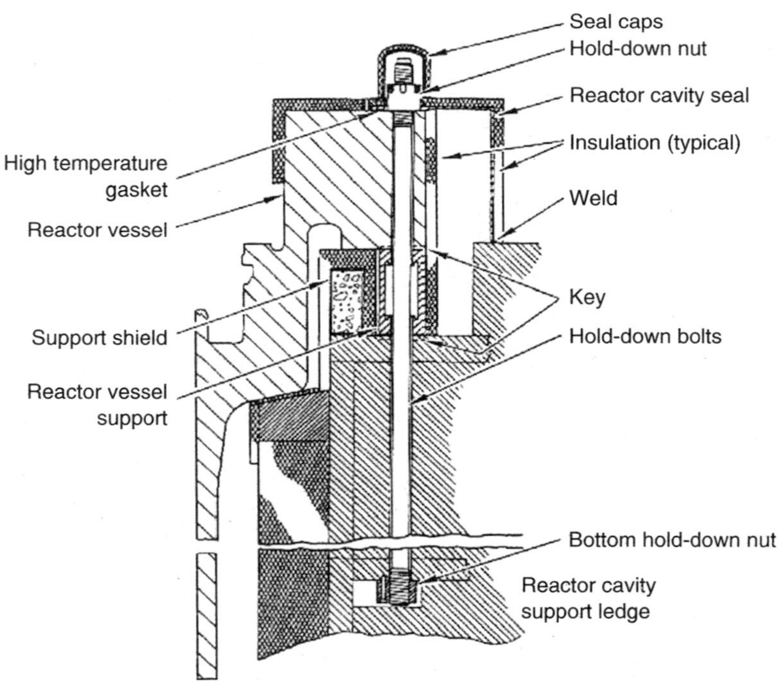

Figure 12.6 Reactor vessel support configuration.

plates welded to two concentric shells of Inconel 600, and operate between the temperatures of 204°C at the reactor vessel flange and 66°C at the ledge. The shells in the box-ring structure provide the insulating capability; that is, they minimize the heat flow from the reactor vessel to the support ledge. The secondary stresses due to temperature distribution and associated fatigue damage in the support structure has been assessed although subsection NF of the ASME Code, Section III does not specifically address this aspect.

The mechanical loads on the reactor vessel support structure include the dead weight and seismic loads. The thermal loads include the differential thermal expansion loads for the normal reactor operation as well as the temperature response of the support structure for a number of postulated thermal transients. These include the reactor closure head heating system overpower, full and partial loss of closure head heaters and reactor vessel heat-up and cool-down events.

The finite-element analysis model of the reactor vessel support structure is shown in Figure 12.7. This is an axisymmetric model and includes a portion of the reactor vessel flange, the box-ring support and an idealization of the support ledge. Due to the nonaxisymmetric nature of the seismic forces and moments the structures are modeled using harmonic axisymmetric finite elements.[1] For these elements the load is defined as a series of harmonic functions (Fourier series). The responses of the structure have been evaluated for unit forces along and unit forces about three orthogonal axes and subsequently scaled up to the actual load magnitudes.

The reactor vessel support has been evaluated for steady-state temperature distribution using a two-dimensional axisymmetric finite-element model (shown in Figure 12.8). This model includes the box-ring structure, a portion of the reactor vessel and a stiffness representation of the support ledge. The model also includes the hold-down bolts. Since the temperature distribution for the normal reactor operation shows no appreciable variation in the azimuthal direction, a two-dimensional model is deemed adequate.

The preload in the hold-down bolts is introduced by means of thermal prestrain. This involves imposing artificial temperature in the bolts sufficiently lower than the assembly temperature to achieve the desired preload. By this method the bolt preload can be monitored very effectively. The temperatures in the reactor vessel support system are obtained using a thermal analysis that employs a model identical to Figure 12.8. The ledge temperature of 66°C is used as a boundary condition in the thermal analysis. The bolt temperature distribution is also evaluated in the thermal analysis. The computed bolt temperature is superimposed onto the fictitious bolt preload temperature. The stresses are evaluated in the stress analysis model using the computed temperatures.

A three-dimensional model extending 180° of the reactor vessel support is employed to study the thermal and structural responses of the assembly for various thermal transients (Figure 12.9). The three-dimensional model

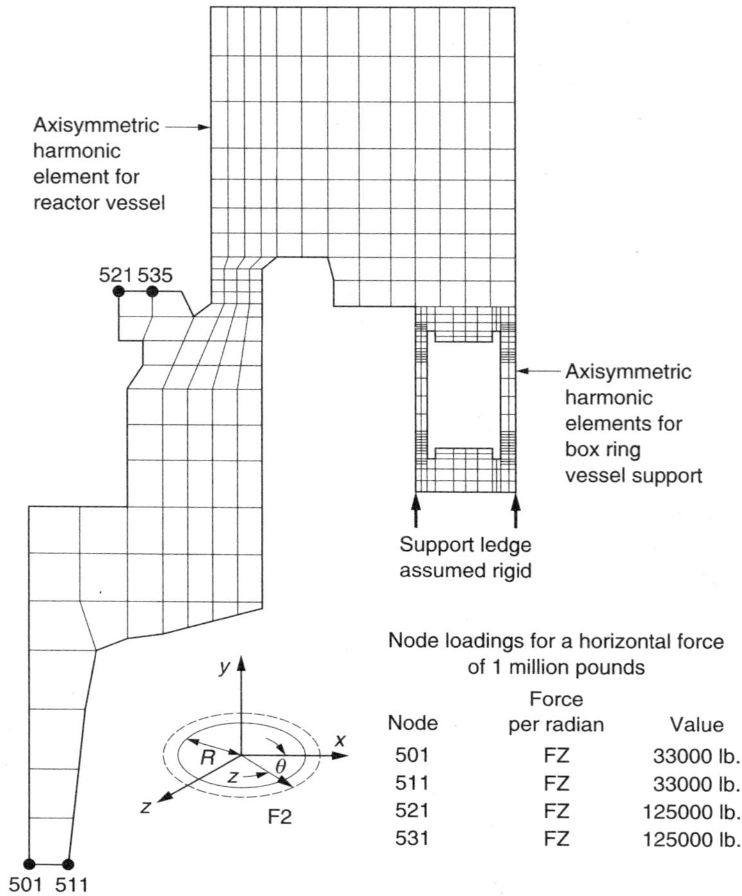

Node loadings for a horizontal force of 1 million pounds

Node	Force per radian	Value
501	FZ	33000 lb.
511	FZ	33000 lb.
521	FZ	125000 lb.
531	FZ	125000 lb.

Figure 12.7 Finite-element model for mechanical load stress analysis.

with 30 elements in the circumferential direction (6° apart) is judged to be sufficiently accurate to pick any nonaxisymmetric effects due to local hotspots as a consequence of the transients. The thermal model simulates the closure head heating system and the hold-down bolts. The thermal boundary condition includes interfaces with the closure head access area, the closure head, the support ledge and the reactor vessel. The worst thermal stress condition in the box-ring support can be defined by the end of the transient (or quasi-steady) condition, eliminating the need for performing a detailed transient thermal analysis. The temperature distribution obtained is used for subsequent stress analysis. The stress analysis model predicts peak stresses at the plate-to-shell junctures in the box-ring support structure. The stress evaluation results are indicated in Table 12.5 along with appropriate allowables. It can be seen that the allowables are

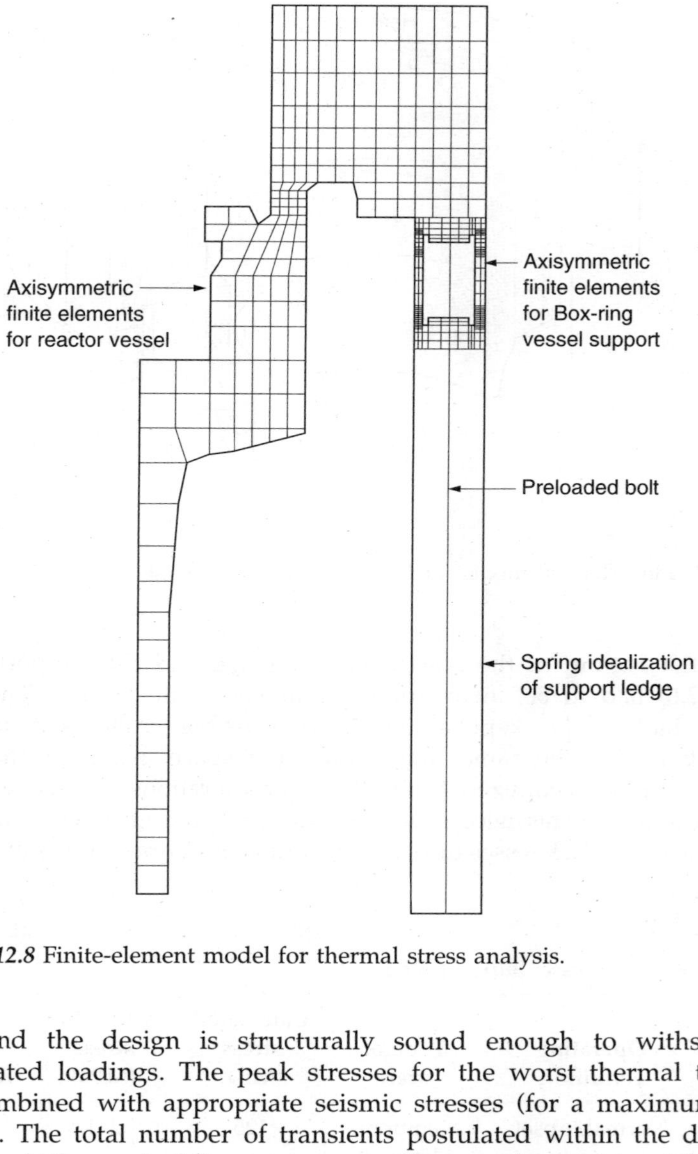

Figure 12.8 Finite-element model for thermal stress analysis.

met and the design is structurally sound enough to withstand the postulated loadings. The peak stresses for the worst thermal transients are combined with appropriate seismic stresses (for a maximum of five cycles). The total number of transients postulated within the design life (assumed 30 years) of the reactor vessel support is considered for fatigue evaluation. The fatigue evaluation summary is indicated in Table 12.6 that shows a fatigue usage of 0.128, thereby demonstrating adequacy for cyclic operation for the design life.

As mentioned earlier, the critical locations for the reactor vessel support structure were the intersections between the Inconel 600 shells and the low alloy-steel plates. These locations were analyzed in detail for maximum stress range and fatigue. The constraints in the design were the thickness of the plates, the thickness of the shell, and the length of the shell. The distance

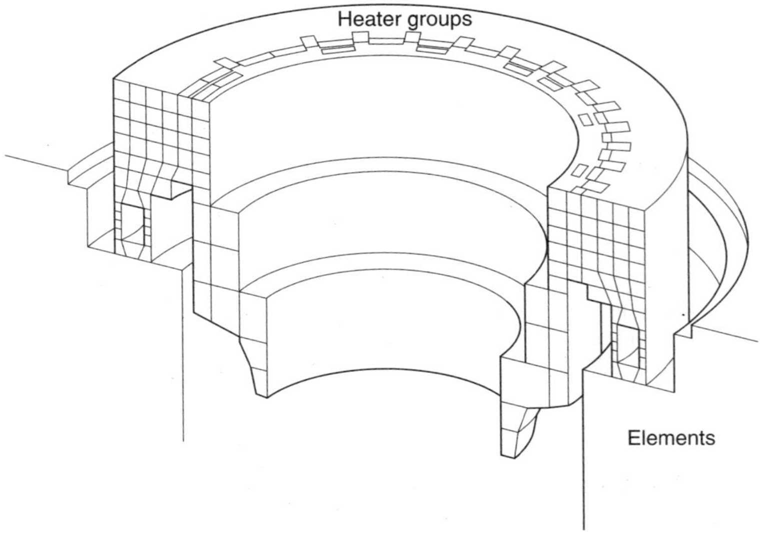

Figure 12.9 Finite-element model for thermal transient analysis.

between the bottom of the reactor vessel flange and the support ledge (Figure 12.6) had to be an invariant parameter in the design. The shell thickness had to be kept as small as possible from heat transfer considerations. At the same time within the space available, the shell length had to be maximized from stress considerations because of large thermal gradients. This necessitated iterating the design over a range of shell lengths and thicknesses until an optimum configuration was obtained.

Table 12.5 Reactor Vessel Support Stresses

Location	Operating condition	Type of stress	Calculated stress (MPa)	Allowable stress (MPa)	Remarks
Shell	Normal/upset	Membrane +bending	125	241	1.5 S_m
Shell	Normal/upset	Primary +secondary	301	482	3.0 S_m
Hold-down bolts	Normal/upset	Membrane	370	392	2.0 S_m
Shell	Emergency (earthquake-safe shutdown)	Membrane +bending	181	185	80% of collapse load

Table 12.6 Reactor Vessel Support Fatigue Evaluation

Transient	Alternating stress (MPa)	Number of cycles (n)	Allowable no. of cycles (N)	Fatigue usage (n/N)
Closure head heat-up	896	5	200	0.025
Full loss of head heaters	606	15	600	0.025
Partial loss of head heaters	619	21	600	0.035
Head heating system overpower	606	21	600	0.035
Reactor vessel heat-up	606	5	600	0.008

Cumulative fatigue usage factor $\Sigma\ (n/N) = 0.128$.

References

1. American Society of Mechanical Engineers, Boiler and Pressure Vessel Code, ASME, New York.
2. Chuse, R., and Eber, S.M., *Pressure Vessels: The ASME Code Simplified*, 6th edn., McGraw-Hill, New York, 1984.
3. Chattopadhyay, S., A comparison of fatigue damage estimates using the ASME Code and local strain approaches, *ASME Paper 84-W A-DE- 7*, ASME, 1984.
4. Chattopadhyay, S., Structural analysis of the reactor vessel support structure for the Clinch River breeder reactor plant, *ASME Paper 82-PVP-36*, ASME, 1982.

Appendix A

Review of solid mechanics

Contents

A.1 Introduction

In order to understand the equations in pressure vessel design, it is important to go over some of the basics of solid mechanics. Although it is essential primarily to learn these concepts developed in the classical theories of plates and shells, for the sake of completeness we shall include the concepts of stress, strain, and constitutive relationships and move over to the theories of elasticity and plasticity.

Before we move on it is important to realize that a body or a structure like that of a pressure vessel is a deformable one. As such the mechanics of such a body would involve deformations being set up as a result of a system of applied forces. The systems of forces that may act on a body are body forces (acting along the entire volume) and surface forces (acting along the external surface of the body). Within the domain of body forces in pressure vessels, we can talk about gravity (dead weight) and seismic loads. As far as the surface forces are concerned, we can mention fluid pressure – which is what a pressure vessel is primarily designed for. Other surface forces might include contact forces due to preload in bolted joints. Under the action of body and surface forces, the body is in equilibrium.

A.2 Concept of stress

If we consider a straight rod of cross-sectional area, A, subjected to a tensile load, F. There the quantity F/A is called the stress, σ. This concept is generalized to the concept of stress at a point. Let us consider a section of a body subject to body forces and surface forces. Figure A.1 shows a cutaway section through the body. This section contains a small area ΔA and is oriented with an outward normal n.

The internal force ΔF is resolved into a normal component ΔF_n and an in-plane component ΔF_i. The normal stress and the shear stress are given by

$$\sigma = \lim_{\Delta A \to 0} \frac{\Delta f_n}{\Delta A}$$

$$\tau = \lim_{\Delta A \to 0} \frac{\Delta f_i}{\Delta A} \tag{A.1}$$

The shear stress may be further resolved into two in-plane components (orthogonal to each other).

Now consider at each point O in the body of Figure A.2, three mutually perpendicular planes, the x-plane (or x_1), the y-plane (or x_2), and the z-plane (or x_3). Across each plane we have a normal component, which for one plane will be denoted as σ_{11} and the two in-plane components of the shear as $\tau_{1,2}$ and $\tau_{1,3}$. Note that the first subscript refers to the plane while the second refers to the direction of the stress component. Note that for the face with outward normal in the negative x_1 direction (this left face) the direction of the components τ_{13} and τ_{12} are reversed from that of the

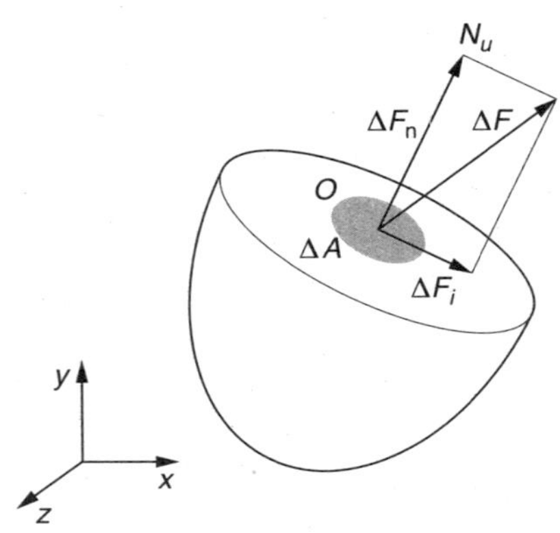

Figure A.1 Normal and shear stresses.

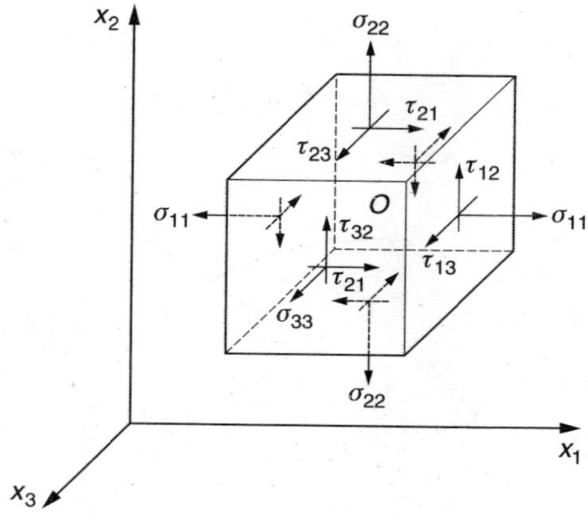

Figure A.2 Stress components.

positive x_1 face. The force equilibrium for the differential parallelepiped element leads to

$$\frac{\partial \sigma_{1,1}}{\partial x_1} + \frac{\partial \tau_{2,1}}{\partial x_2} + \frac{\partial \tau_{3,1}}{\partial x_3} = 0$$

$$\frac{\partial \tau_{1,2}}{\partial x_1} + \frac{\partial \sigma_{2,2}}{\partial x_2} + \frac{\partial \tau_{3,2}}{\partial x_2} = 0$$

$$\frac{\partial \tau_{1,3}}{\partial x_1} + \frac{\partial \tau_{2,3}}{\partial x_2} + \frac{\partial \sigma_{3,3}}{\partial x_3} = 0 \qquad \text{(A.3)}$$

The moment equilibrium leads to $\tau_{1,2} = \tau_{2,1}$, $\tau_{1,3} = \tau_{3,1}$, $\tau_{2,3} = \tau_{3,2}$.

A.3 Equations of equilibrium in a cylindrical system

Figure A.3 shows a differential element of a rotationally symmetric three-dimensional body in a cylindrical coordinate system.

The stress tensor in cylindrical coordinate system is given by

$$\begin{bmatrix} \sigma_r & \tau_{r\theta} & \tau_{rz} \\ \tau_{\theta r} & \sigma_\theta & \tau_{\theta z} \\ \tau_{zr} & \tau_{z\theta} & \sigma_z \end{bmatrix} \qquad \text{(A.5)}$$

Force equilibrium in the r, θ, and z directions gives

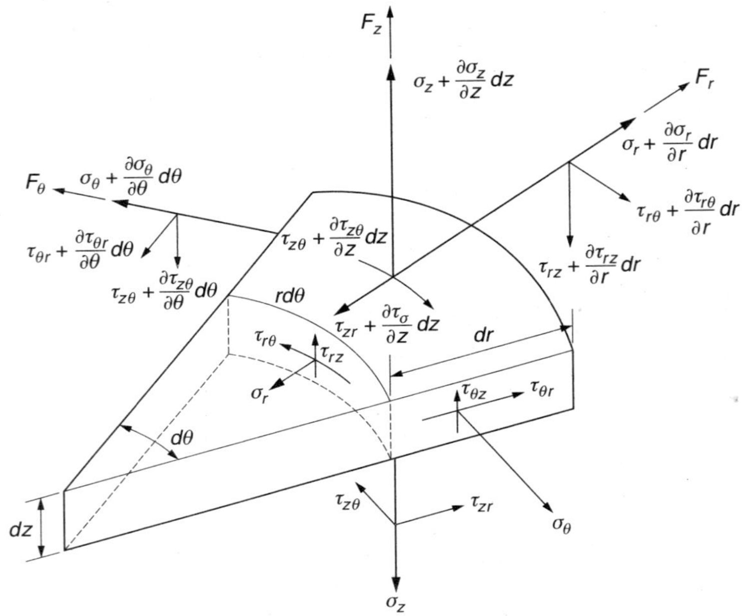

Figure A.3 Stress components in a rotationally symmetric body.

$$\frac{\partial \sigma_r}{\partial r} + \frac{\partial \tau_z}{\partial z} + \frac{1}{r}\frac{\partial \tau_{r\theta}}{\partial \theta} + \frac{\sigma_r - \sigma_\theta}{r} = 0$$

$$\frac{\partial \tau_{r\theta}}{\partial r} + \frac{1}{r}\frac{\partial \tau_{r\theta}}{\partial \theta} + \frac{\partial \tau_{\theta z}}{\partial z} + \frac{2\tau_{r\theta}}{r} = 0$$

$$\frac{\partial \tau_{rz}}{\partial r} + \frac{1}{r}\frac{\partial \tau_{z\theta}}{\partial \theta} + \frac{\partial \tau_{\theta z}}{\partial z} + \frac{2\sigma_z}{\partial z} + \frac{\tau_{rz}}{r} = 0 \qquad \text{(A.6)}$$

For situations involving axial symmetry, $\tau_{r\theta} = \tau_{\theta z} = 0$ and $\tau_{\theta z} = \tau_{z\theta} = 0$ and there is no dependence on θ. This gives

$$\frac{\partial \sigma_r}{\partial r} + \frac{\partial \tau_{zr}}{\partial z} + \frac{\sigma_r - \sigma_\theta}{r} = 0$$

$$\frac{\partial \tau_{rz}}{\partial r} + \frac{\partial \sigma_z}{\partial z} + \frac{\tau_{rz}}{r} = 0 \qquad \text{(A.7)}$$

A.4 Principal stresses

For any plane passing through a point the stress resultant in general possesses a normal stress component and two shear stress components. We would like to investigate whether there are planes for which there are no shear stresses. Such planes are called principal planes, and the normal stresses along those planes are called principal stresses.

The three principal stresses can be found by solving the following cubic equation:

$$\sigma^3 - I\sigma^2 + II\sigma - III = 0 \tag{A.8}$$

where

$$I = -\sigma_{1,1} + \sigma_{2,2} + \sigma_{3,3}$$

$$II = (\sigma_{1,1}\sigma_{2,2} + \sigma_{2,2}\sigma_{3,3} + \sigma_{3,3}\sigma_{1,1}) - \tau_{1,2}^2 - \tau_{2,3}^2 - \tau_{3,1}^2$$

$$III = \begin{Vmatrix} \sigma_{1,1} & \tau_{1,2} & \tau_{1,3} \\ \tau_{2,1} & \sigma_{2,2} & \tau_{2,3} \\ \tau_{3,1} & \tau_{3,2} & \sigma_{2,3} \end{Vmatrix} \tag{A.9}$$

I, II, and *III* are stress invariants, because their values do not change with the rotations of the frame of reference. If we derive the roots of the Eq. (A.8) by $\sigma_1, \sigma_2, \sigma_3$, then

$$I = \sigma_1 + \sigma_2 + \sigma_3$$

$$II = \sigma_1\sigma_2 + \sigma_2\sigma_3 + \sigma_3\sigma_1$$

$$III = \sigma_1\sigma_2\sigma_3 \tag{A.10}$$

A.5 Strain

If we consider a straight rod of length l and when a load is applied the length is increased by Δl. Then the ratio of $\Delta l/l$ is called the strain Δ. Looking at a two-dimensional deformation pattern, for a rectangle *ABCD*, is shown in Figure A.4. The two dimensional strain components are defined as

$$\varepsilon_{1,1} = \frac{\partial u}{\partial x_1}$$

$$\varepsilon_{2,2} = \frac{\partial v}{\partial x_2}$$

$$\gamma_{1,2} = 2\varepsilon_{1,2} = \left(\frac{\partial u}{\partial x_2} + \frac{\partial v}{\partial x_1}\right) \tag{A.11}$$

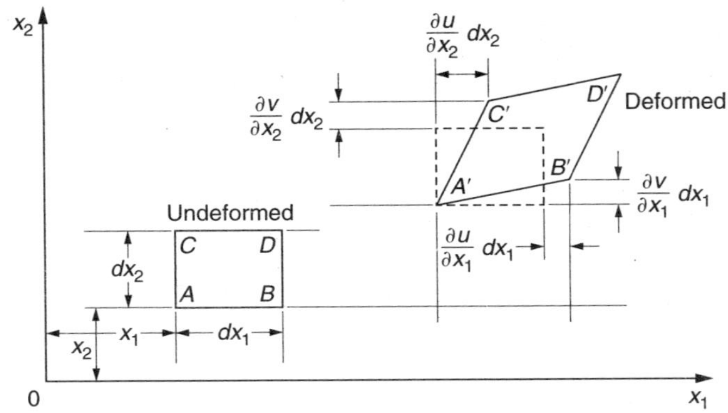

Figure A.4 Strain components in two dimensions.

Generalizing the concept to the three-dimensional case, where u, v, and w are the displacements in x, y, and z directions, the strain tensor components can be represented as

$$
\begin{aligned}
\varepsilon_{1,1} &= \frac{\partial u}{\partial x_1} \quad & \varepsilon_{1,2} &= \frac{1}{2}\left(\frac{\partial u}{\partial x_2} + \frac{\partial v}{\partial x_1}\right) \\
\varepsilon_{1,1} &= \frac{\partial v}{\partial x2} \quad & \varepsilon_{1,1} &= \frac{1}{2}\left(\frac{\partial u}{\partial x_3} + \frac{\partial w}{\partial x_1}\right) \\
\varepsilon_{1,1} &= \frac{\partial w}{\partial x3} \quad & \varepsilon_{1,1} &= \frac{1}{2}\left(\frac{\partial v}{\partial x_3} + \frac{\partial w}{\partial x_2}\right)
\end{aligned}
\tag{A.12}
$$

In a cylindrical coordinate system the strain components (or strain displacement relationships) as given by (where u_r, u_θ, u_z are displacements in r, θ, and z directions):

$$
\begin{aligned}
\varepsilon_r &= \frac{\partial u_r}{\partial r} \quad & \gamma_{r\theta} &= \frac{1}{r}\left(\frac{\partial u_r}{\partial \theta} + \frac{\partial u_\theta}{\partial r} - \frac{u_\theta}{r}\right) \\
\varepsilon_\theta &= \frac{u_r}{r} \quad & \gamma_{\theta z} &= \frac{\partial u_\theta}{\partial z} + \frac{1}{r}\frac{\partial u_z}{\partial \theta} \\
\varepsilon_z &= \frac{\partial u_z}{\partial z} \quad & \gamma_{zr} &= \frac{\partial u_r}{\partial z} + \frac{\partial u_z}{\partial r}
\end{aligned}
\tag{A.13}
$$

For an axially symmetric case, we have $u_\theta = 0$ and no dependence on θ. This gives

$$\varepsilon_r = \frac{\partial u_r}{\partial r} \qquad \gamma_{r\vartheta} = 0$$

$$\varepsilon_\theta = \frac{u_r}{r} \qquad \gamma_{\vartheta z} = 0$$

$$\varepsilon_z = \frac{\partial w}{\partial z} \qquad \gamma_{rz} = \frac{\partial u_r}{\partial z} + \frac{\partial u_z}{\partial r} \tag{A.14}$$

Similar to stresses, there are planes along which there are only normal strains, and no shear strains. Identical relationships such as those derived for principal stresses in Eqs. (A.8), (A.9), and (A.10) can be obtained for principal strains.

A.6 Stress–strain relations

For a three-dimensional state of stress, the strain components are related to stress components by the following relationships. Note these are for the ideal case of a linear elastic material with isotropy and homogeneity, for which two material constraints are needed: E, the elastic modulus, and ν, Poisson's ratio.

$$\varepsilon_{1,1} = \frac{1}{E}\left[\sigma_{1,1} - \nu\left(\sigma_{2,2} + \sigma_{3,3}\right)\right]$$

$$\varepsilon_{2,2} = \frac{1}{E}\left[\sigma_{2,2} - \nu\left(\sigma_{1,1} + \sigma_{3,3}\right)\right]$$

$$\varepsilon_{3,3} = \frac{1}{E}\left[\sigma_{3,3} - \nu\left(\sigma_{1,1} + \sigma_{2,2}\right)\right]$$

$$\varepsilon_{1,2} = \frac{1+\nu}{E}\tau_{1,2}$$

$$\varepsilon_{2,3} = \frac{1+\nu}{E}\tau_{2,3}$$

$$\varepsilon_{1,2} = \frac{1+\nu}{E}\tau_{3,1} \tag{A.15}$$

A.7 Elastic plane problems

This approach is relevant to structures commonly used in pressure vessel construction. What this means is that the problem can be treated as a two-dimensional situation, wherein the stress or the strain components at every point in the body are functions only of the reference coordinates parallel to that plane. For example, a long, thick-walled cylinder subjected to internal pressure can be treated as a plane elastic problem.

Plane elastic problems can be formulated either as a "plane stress" or a "plane strain" one. For either one, the solutions are usually developed in terms of a stress function.

A.7.1 Plane strain

A state of plane strain exists with respect to the x–y plane when ε_x, γ_{yz}, and γ_{zx} are identically zero and the remaining strain components are functions of x and y only and are

$$\varepsilon_x = \frac{\partial u}{\partial x}, \ \varepsilon_y = \frac{\partial u}{\partial y} \text{ and } \gamma_{xy} = \frac{\partial v}{\partial x} + \frac{\partial u}{\partial y} \tag{A.16}$$

It may be shown further that $w = 0$, and u and v are functions of x and y alone.

The equilibrium equations from (A.3) are then given by

$$\frac{\partial \sigma_x}{\partial x} + \frac{\partial \tau_{xy}}{\partial y} = 0$$

$$\frac{\partial \sigma_{xy}}{\partial x} + \frac{\partial \sigma_y}{\partial y} = 0 \tag{A.17}$$

and σ_x, σ_y, and τ_{xy} are functions of x and y only.

The stress–strain relationship using Eq. (A.15):

$$\varepsilon_x = \frac{1}{E}\left[\sigma_x - \nu\left(\sigma_y + \sigma_z\right)\right] \tag{A.18a}$$

$$\varepsilon_x = \frac{1}{E}\left[\sigma_y - \nu(\sigma_x + \sigma_z)\right] \tag{A.18b}$$

$$\varepsilon_x = \frac{1}{E}\left[\sigma_x - \nu\left(\sigma_y + \sigma_z\right)\right] \tag{A.18c}$$

$$\gamma_{xy} = 2\varepsilon_{xy} = \frac{2(1+\nu)}{E}\tau_{xy} \tag{A.18d}$$

$$\gamma_{xy} = 2\varepsilon_{yz} = \frac{2(1+\nu)}{E}\tau_{yz} \tag{A.18e}$$

$$\gamma_{zx} = 2\varepsilon_{zz} = \frac{2(1+\nu)}{E}\tau_{zx} \tag{A.18f}$$

Note that $\varepsilon_z = 0$, $\gamma_{yz} = 0$ and $\gamma_{zx} = 0$. From Eq. (A.18c) therefore, we have

$$\sigma_x = \gamma\left(\sigma_y + \sigma_z\right)$$

Substituting the above in Eqs. (A.18a) and (A.18b) we have

$$\varepsilon_x = \frac{1+\nu}{E}\left[\sigma_x - \nu\left(\sigma_x + \sigma_y\right)\right]$$

$$\varepsilon_y = \frac{1+\nu}{E}\left[\sigma_y - \nu\left(\sigma_x + \sigma_y\right)\right]$$

$$\gamma_{xy} = \frac{2(1+\nu)}{E}\tau_{xy} \tag{A.19}$$

A.7.2 Plane stress

A state of plane stress exists with respect to x–y plane when σ_x, τ_{yz} and τ_{yz} are zero, and σ_x, σ_y and τ_{xy} are functions of x and y only. The equilibrium equations are given by (A.17) for the x and y directions. In this case, however, σ_x is zero, but σ_z is nonzero. The strain displacement relations are the same as Eq. (A.17) with the additional relationship

$$\varepsilon_z = \frac{\partial w}{\partial z} \tag{A.20}$$

Further analysis leads to the important result that ε_z is a linear function of x and y only.

It turns out that the plane stress assumption, wherein σ_x, σ_y and τ_{xy} are independent of z, is good approximation for thin plates subjected to forces that are uniformly distributed over the thickness of the boundary and are parallel to the middle plane.

The stress–strain relations for plane stress are given by

$$\left.\begin{aligned}
\varepsilon_x &= \frac{1}{E}\left(\sigma_x - \nu\sigma_y\right)\\
\varepsilon_y &= \frac{1}{E}\left(\sigma_y - \nu\sigma_x\right)\\
\varepsilon_z &= \frac{-\upsilon}{E}\left(\sigma_x + \nu\sigma_y\right)\\
\gamma_{xy} &= \frac{-\upsilon}{E}\left(\sigma_x + \nu\sigma_y\right)
\end{aligned}\right\} \tag{A.21}$$

A.7.3 Stress function formulation

Manipulating the equilibrium equations and stress displacement equations, we have for the case of the plane strain (with the notation: $\nabla^2 \equiv \frac{\partial^2}{\partial x^2} + \frac{\partial^2}{\partial y^2}$):

$$\nabla^2 u + \frac{1}{(1-2\upsilon)}\frac{\partial}{\partial x}\left(\frac{\partial u}{\partial x}+\frac{\partial v}{\partial y}\right) = 0$$

$$\nabla^2 v + \frac{1}{(1-2\upsilon)}\frac{\partial}{\partial y}\left(\frac{\partial u}{\partial x}+\frac{\partial v}{\partial y}\right) = 0 \tag{A.22}$$

and for the case of plane stress

$$\nabla^2 u + \frac{1+\nu}{(1-\upsilon)}\frac{\partial}{\partial x}\left(\frac{\partial u}{\partial x}+\frac{\partial v}{\partial y}\right) = 0$$

$$\nabla^2 v + \frac{1}{(1-2\upsilon)}\frac{\partial}{\partial y}\left(\frac{\partial u}{\partial x}+\frac{\partial v}{\partial y}\right) = 0 \tag{A.23}$$

Eqs. (A.22) and (A.23) are known as Navier's equations. We also observe that replacing υ by $\upsilon/(1-\upsilon)$ in the plane stress, Eqs. (A.23) are obtained.

It can be rigorously shown using Eqs. (A.22) and (A.23) that for both plane stress and plane strain problems, the following relations hold in the absence of body forces:

$$\nabla^2(\sigma_x + \sigma_y) = 0 \tag{A.24}$$

If the stresses can be written down in terms of $\Phi\ (x, y)$, also known as the Airy stress function, such that

$$\sigma_x = \frac{\partial^2 \phi}{\partial y^2}$$

$$\sigma_y = \frac{\partial^2 \phi}{\partial x^2}$$

$$\tau_{xy} = -\frac{\partial^2 \phi}{\partial x \partial y} \tag{A.25}$$

Eq. (A.24) then reduces to

$$\nabla^2(\nabla^2 \phi) = 0$$

or

$$\nabla^4 \phi = 0 \tag{A.26}$$

where the bi-harmonic operator is given by

$$\nabla^4 \equiv \frac{\partial^4}{\partial x^4} + 2\frac{\partial^2}{\partial x^2 \partial y^2} + \frac{\partial^4}{\partial y^4}$$

A.8 Plasticity

For a bar loaded in tension the stress generally is linear with strain representing a linear elastic behavior. When the load is removed, the bar comes back to its original shape without any permanent deformation, thus depicting an elastic response of the material. Beyond the linear portion of the stress–strain curve or beyond the proportional limit (typically below the yield point) the stress–strain relationship is generally nonlinear and the material progressively moves to the plastic region characterized by a permanent deformation. It is conceivable to have an elastic material having nonlinear stress–strain behavior as long as there is no permanent deformation upon unloading.

Typically the stress for which a permanent strain of 0.002 or 0.2 percent occurs upon the removal of load is defined as the yield strength. This is also referred to as 0.2 percent offset yield strength. This is a standard means of characterizing yield behavior, since quite a few materials do not have a well-defined material yield, where a marked change from predominantly elastic to plastic behavior takes place. As the load is progressively increased for the bar in tension beyond the yield, some materials display strain-hardening behavior before failure. For some brittle materials, such as cast iron, the bar in tension breaks without any significant strain hardening. In either case, the nominal stress at which the material breaks is known as the ultimate tensile strength, which is the load causing fracture divided by the original cross-section area. However, the cross-sectional area in most materials is reduced as the applied load increases. Therefore the true stress is often larger than the nominal stress, which is based on the original cross-sectional area. The true stress at fracture is thus considerably higher than the nominal ultimate tensile strength.

A typical stress–strain curve is shown in Figure A.5. The proportional limit is indicated as S_p and the yield strength (0.2 percent offset) by S_y; S_u denotes the ultimate strength. It is clear from Figure A.5 that for

$$\sigma \leq S_\mathrm{p}, \;\; \varepsilon = \frac{\sigma}{E} \tag{A.27}$$

and for

$$\sigma > S_\mathrm{p}, \;\; \varepsilon = \frac{\sigma}{E} + K\sigma^{\frac{1}{n}} \tag{A.28}$$

Here, E is the slope of the line shown in dotted lines in Figure A.5, and referred to as the modulus of elasticity or Young's modulus. The index n is

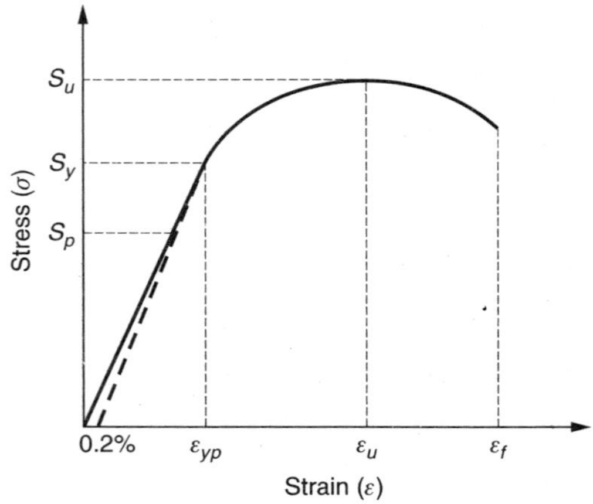

Figure A.5 Stress vs. strain.

known as the strain hardening exponent and K (as is n) is a material constant. The behavior represented by Eq. (A.28) is the so-called Ramberg Osgood relationship. In pressure vessel design, the strain-hardening exponent is used within the context of simplified inelastic analysis as indicated in Chapter 9.

Thus far, the discussion has centered on uniaxial tension. Let us see what happens when the bar in tension is unloaded first and then loaded in compression. The typical stress–strain curve is represented in Figure A.6. The loading is initially in tension up to a stress of S_1 at which the unloading process starts and follows a straight line parallel to the elastic curve. Upon unloading as the strain reaches zero, a tensile strain of ε_p remains. Subsequently as the material is loaded in compression, it may yield in

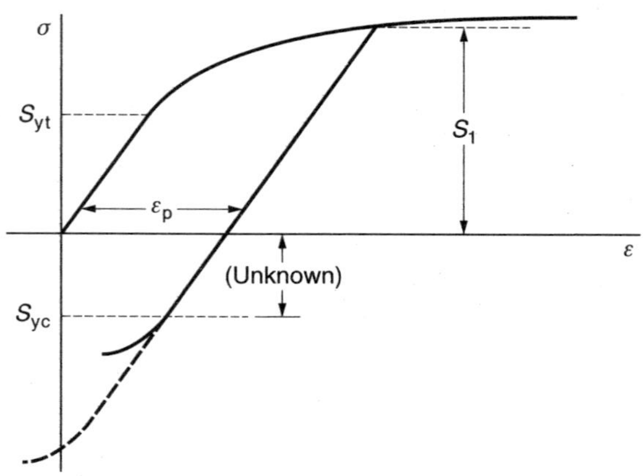

Figure A.6 Bauschinger effect.

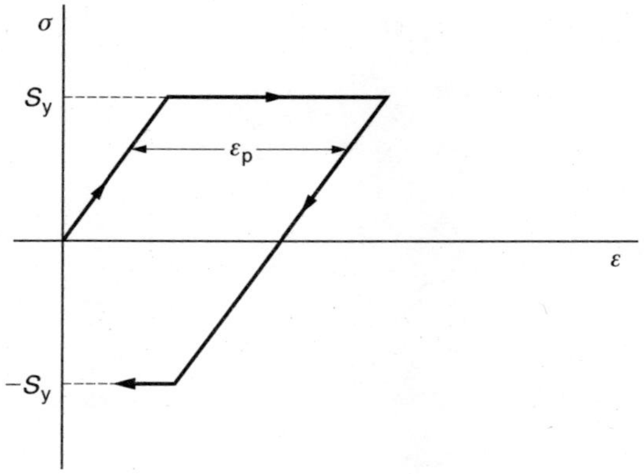

Figure A.7 Elastic–perfect plastic representation.

compression at a compressive stress of S_{yc}. For some materials this compressive yield strength is lower than the tensile yield strength, a phenomenon known as the Bauschinger effect. Moreover, the value of S_{yc} is not clearly defined for a lot of materials. Therefore a concept known as isotropic hardening is used wherein it is assumed that the yield stress in compression is the same as that in tension.

The material behavior model that is generally used for pressure vessel design is the elastic–perfectly plastic representation, shown in Figure A.7. In this model, the assumption of zero strain hardening is employed. This implies that the material has a lower load carrying capacity than one in which strain hardening is present. In the simplified model, the proportional limit, the yield strength, as well as the ultimate strength are identical. This is of course a very conservative assumption.

Appendix B

Review of fatigue and fracture mechanics

Contents

B.1 S–N curves

These curves form the basis for the stress-life approach to fatigue, and are plots of alternating stress, *S*, versus cycles to failure, *N*. The most common procedure for generating the *S–N* curve is the R.R. Moore bending test which uses four-point loading in which a constant moment is applied to a cylindrical specimen (see Figure B.1). This loading produces a completely reversed uniaxial bending stress distribution across the cross-section.

The number of cycles causing fatigue failure depends on the strain range incurred during each load reversal. The critical strain amplitude at which the material can be cycled indefinitely without producing fatigue failure is the endurance limit strain amplitude. This value when multiplied by the elastic modulus, *E*, gives the endurance limit. Figure B.2 shows the best-fit fatigue austenitic steels or nickel–chromium–iron alloys (typical pressure vessel materials). For these materials, the design fatigue (*S–N*) curve is defined over a cyclic range of 10 to 10^6 cycles. Since the austenitic stainless steels and nickel–chromium–iron alloys have no clearly defined endurance limits, the design curves are extended to 10^{11} cycles as shown in Figure B.3. In Figure B.3, the curves *A* and *B* are applicable when the stress intensity ranges are below 187.5 MPa (Curve *B* incorporates the mean stress

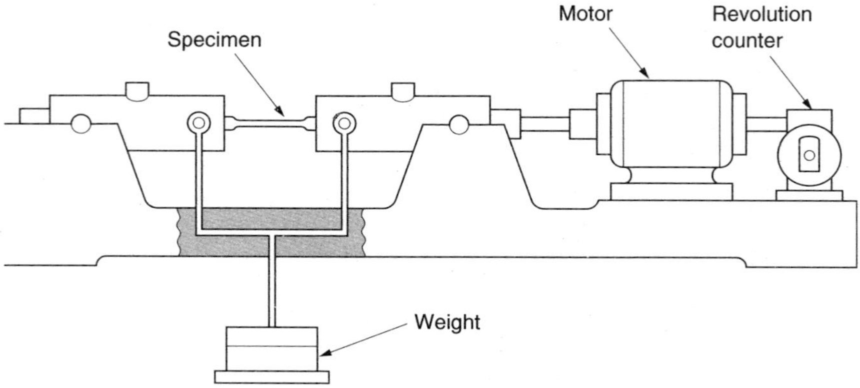

Figure B.1 R.R. Moore bending fatigue test.

correction). Curve C applies when the stress intensity range exceeds 187.5 MPa.

The fatigue evaluation procedure is outlined in Chapter 8 in which it was mentioned how the alternating stress intensity is calculated for the general multiaxial stress state in a pressure vessel component. In addition, the effects of the so-called local structural discontinuities must be evaluated using stress concentration factors determined from theoretical, numerical or numerical techniques. These are referred to as the fatigue strength reduction factors, which generally should not exceed a value of 5.

To account for the effect of elastic modulus (in case it is different to what is defined on the design fatigue curve), the computed alternating stress is multiplied by the ratio of the modulus of elasticity given on the design fatigue curve, to the value of the modulus of elasticity used in the analysis. This value is then entered in the applicable design fatigue curve on the ordinate axis to find the corresponding number of cycles on the abscissa.

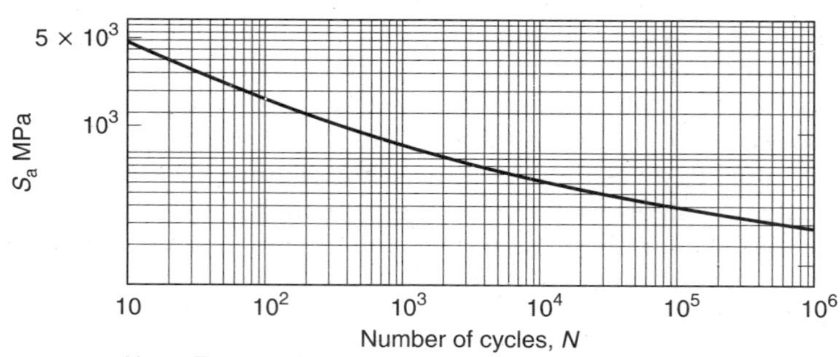

Figure B.2 Fatigue curve for austenitic steels and Ni–Cr–Fe alloys. (Modified from ASME Boiler and Pressure Vessel Code, ASME, New York.)

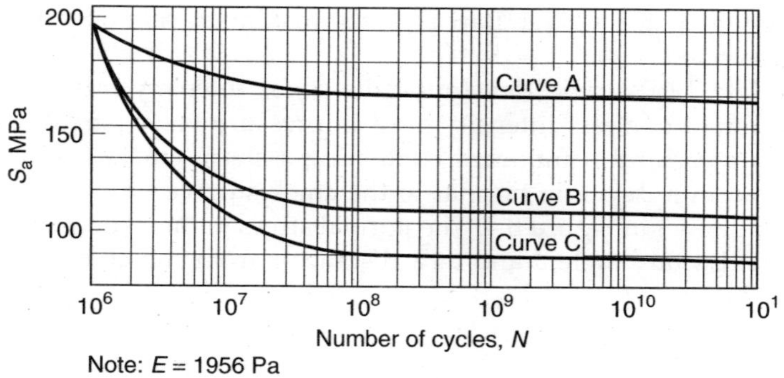

Figure B.3 Extension of fatigue curves beyond 10^6 cycles.
(Modified from ASME Boiler and Pressure Vessel Code, ASME, New York.)

B.2 Cumulative fatigue damage

The linear damage rule is used for the design of pressure vessels for fatigue evaluation. This rule was first proposed by Palmgren and further developed by Miner, and is popularly known as Miner's rule.

The incremental fatigue damage at an alternating stress intensity range S_i equals n_i/N_i, where n_i is the number of operation at this stress range and N_i is the cycles to failure for the alternating stress intensity range S_i. According to Miner's rule the failure takes place when the cumulative damage, U, equals

$$U = \sum_{i=1}^{n} \frac{n_i}{N_i} = 1 \tag{B.1}$$

where the summation has been carried out for m different alternating stress intensity ranges ($i = 1, m$). When U is less than unity the structure is safe from fatigue.

B.3 Basic fracture mechanics

Fracture mechanics is a systems approach to estimate the relationship between the stresses, the flaw geometry, and the material properties as it relates to the structural integrity of the component under consideration. The primary objectives are to prevent failures, promote effective design, and make efficient use of materials. The modern approach seeks a parameter that is a measure of the material's toughness that is independent of the geometry, and can be used with stress analysis to predict fracture loads and critical crack sizes. Using this toughness value from a test specimen in the laboratory, the flaw sizes at which fracture will occur for the structural component used in service can be predicted. The reverse scenario is also applicable where, for a given flaw size, the maximum safe operating stress may be predicted.

In the late 1950s Irwin developed the stress intensity factor approach (this is different to stress intensity used in pressure vessel stress analysis). Consider a structural component containing a sharp crack, subjected to a load applied in a direction normal to the crack surface (known as Mode I loading) as shown in Figure B.4. The normal stress in the y direction, σ_y, at a point located at an angle θ and at a distance r from the crack tip, can be expressed as

$$\sigma_y = \left(\frac{K_I}{\sqrt{2\pi r}}\right) f(\theta) \tag{B.2}$$

where $f(\theta)$ is a trigonometric function of the angle θ. The parameter K_I is a measure of the magnitude of the stress field in the vicinity of the crack tip and is referred to as the Mode I stress intensity factor. This parameter describes the extent of stress intensification resulting from the flaw, and the expressions for its value can be found in various handbooks. Some common load and crack geometries and the corresponding expressions for the stress intensity factors are shown in Figure B.5. As shown in Figure B.5a, for a plate under a uniform tensile stress containing a central crack, the stress intensity factor, K_I is given by

$$K_I = \sigma\sqrt{\pi a} \tag{B.3}$$

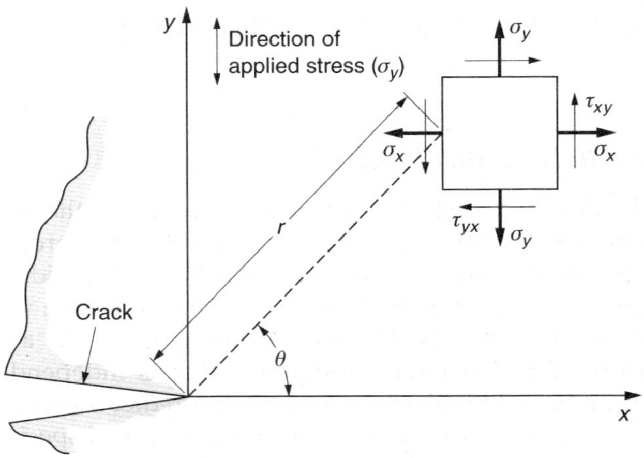

Figure B.4 Mode-I loading of a crack.

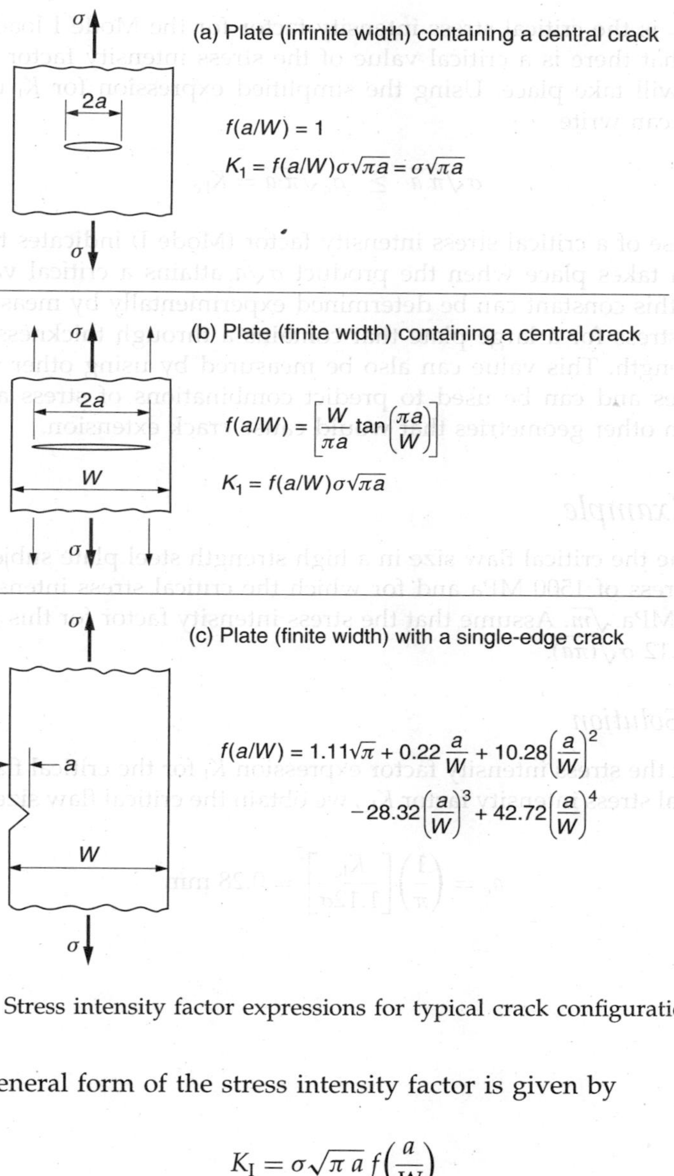

Figure B.5 Stress intensity factor expressions for typical crack configurations.

The general form of the stress intensity factor is given by

$$K_{\mathrm{I}} = \sigma\sqrt{\pi a}\, f\left(\frac{a}{W}\right) \tag{B.4}$$

where $f(a/W)$ is a dimensionless parameter that depends on the geometries of the specimen and the crack. The direct relationship between K_{I} and the stresses and strains at the crack tip leads to the failure criterion,

$$K_{\mathrm{I}} \to K_{\mathrm{Ic}} \text{ at fracture} \tag{B.5}$$

where K_{Ic} is the critical stress intensity factor for the Mode I loading, and implies that there is a critical value of the stress intensity factor at which fracture will take place. Using the simplified expression for K_I using Eq. (B.3), we can write

$$\sigma\sqrt{\pi a} \geq \sigma_c\sqrt{\pi a} = K_{Ic} \tag{B.6}$$

The use of a critical stress intensity factor (Mode I) indicates that crack extension takes place when the product $\sigma\sqrt{a}$ attains a critical value. The value of this constant can be determined experimentally by measuring the fracture stress for a large plate that contains a through thickness crack of known length. This value can also be measured by using other specimen geometries and can be used to predict combinations of stress and crack lengths in other geometries that would cause crack extension.

B.4 Example

Determine the critical flaw size in a high strength steel plate subjected to a tensile stress of 1500 MPa and for which the critical stress intensity factor $K_{Ic} = 50$ MPa $\sqrt{m}$. Assume that the stress intensity factor for this geometry is $K_I = 1.12\ \sigma\sqrt{(\pi a)}$.

B.4.1 Solution

Equating the stress intensity factor expression K_I for the critical flaw size to the critical stress intensity factor K_{Ic}, we obtain the critical flaw size, a_c. Thus

$$a_c = \left(\frac{1}{\pi}\right)\left[\frac{K_{Ic}}{1.12\sigma}\right]^2 = 0.28 \text{ mm} \tag{B.7}$$

Appendix C

Limit analysis

Contents

C.1 Shape factor

An important concept in the area of structural stability of a cross-section resisting bending is that of the shape factor. Basically the shape factor of a cross-section in a beam bending mode is the ratio of the collapse moment to the maximum elastic moment.

Let us consider the case of a beam of rectangular cross-section of width b and thickness h subjected to a bending moment M as shown in Figure C.1. As we increase the moment from zero, the limit stress S_y is reached at the outer surface. With a further increase in moment plastic regions extend from the outside to the inside. As the moment becomes sufficiently large, the beam fails in plastic collapse. If we denote the value of the moment when the limit stress is reached as M_o, then

$$S_y = \frac{6M_o}{bh^2}$$

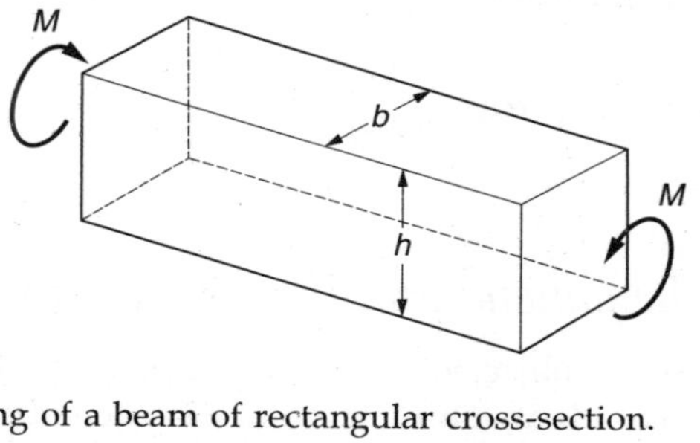

Figure C.1 Bending of a beam of rectangular cross-section.

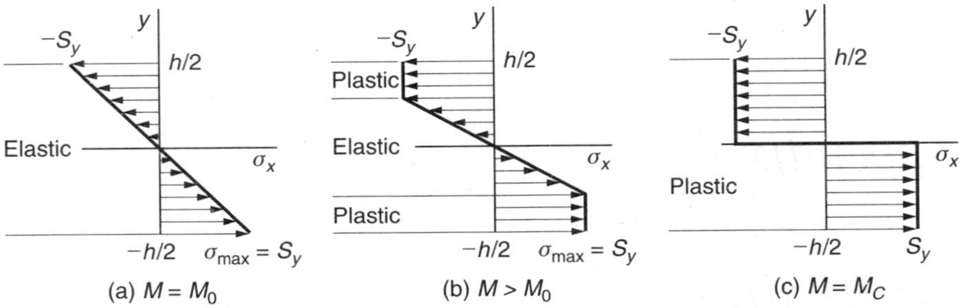

Figure C.2 Elastic–plastic and fully plastic bending of a beam.

or

$$M_o = \frac{bh^2 S_y}{6} \tag{C.1}$$

This moment M_o is defined as the maximum elastic moment producing the limit stress S_y at the outer surface of the beam.

Figure C.2 shows the three different stages of the response of the beam to the applied bending moment. Figure C.2a shows the stress distribution in the beam for the case of the elastic moment M_o when the outer surface has just yielded. There is a linear distribution of the bending stress throughout the cross-section. Figure C.2b shows that the yielding has progressed to a certain depth inside the beam. The rest of the beam stays elastic and has a linear distribution of bending stress. Finally, Figure C.2c shows that the entire cross-section becomes plastic. If we denote the moment for this case as M_c, then

$$M_c = \int_{y=-h/2}^{h/2} \sigma\, y\, dy = bS_y(h/2)(h/2) = S_y \frac{bh^2}{4} \tag{C.2}$$

From Eqs. (C.1) and (C.2), the shape factor of a beam of rectangular cross-section, α_{rect} is determined as

$$\alpha_{rect} = \frac{M_c}{M_o} = 1.5 \tag{C.3}$$

C.2 Collapse phenomena and limit theorems

When a pressure vessel is subjected to an internal pressure and the pressure is monotonically increased, an ultimate value is reached when the vessel collapses. The plasticity developed in the section is completely tensile and a plastic hinge forms before the collapse occurs. When the primary mean stresses are sufficiently large to produce collapse, a uniform tensile plasticity is developed at all points within the cross-section of the vessel. This typically represents collapse due to primary membrane stress intensity, P_m.

It is possible to have collapse due to loads that produce large bending moments over the full cross-section of the vessel. The mode of collapse is bending and the collapse takes place when there is a full plastic hinge developed over the cross-section. In this pattern of plasticity, part of the section is fully plastic in tension, and the remainder of the section is fully plastic in compression. The set of bending stresses generated by sustained bending is the primary bending stress intensity, P_b. When the primary bending stress intensity P_b reaches the limit stress (typically the yield strength) at the surface (at one or both), the structure does not collapse. It is able to carry additional load until the entire cross-section becomes plastic. Typically for a rectangular cross-section, the moment required to produce collapse is 150 percent of the moment required to initiate plastic deformation at the outer surface (Eq. C.3) and this is recognized by pressure vessel design codes worldwide.

We can contrast this situation to what happens when the stress distribution is such that no gross collapse takes place. Typically, the average value across the section would correspond to the primary membrane stress intensity, P_m. The linear distribution would correspond to P_b, the primary bending stress and the remaining stress would possibly correspond to secondary stress, Q. The stress distribution characterized by Q cannot produce gross collapse but could cause large deformations. We can of course have secondary stresses due to boundary constraints, and at a junction of two different cross-sections, for example at the shell to head or shell to flange junctions. Secondary stresses could also be produced by thermal stresses which are self-limiting in nature.

For structures with complex geometries, the exact collapse loads may be fairly difficult to compute. The collapse analyses are based on theorems that establish lower and upper bounds of the collapse load. The collapse load is somewhere in between the upper and lower bound loads. Thus a conservative estimate of the collapse load is provided by the lower bound limit load.

The upper bound theorem states:

> When the work done by the external loads is equal to, or greater than, the internal work done by deformation in the fully plastic regions of a structure, the external loads will represent an upper bound to the collapse load of a structure.

The lower bound theorem states:

> If any internal stress distribution can be found which does not exceed the yield stress at any point, and which is in equilibrium with the applied sustained loading, then the structure does not collapse.

Another way of stating the upper bound theorem is that during collapse, the work done by the external loads is equal to the work dissipated by the internal loads. The lower bound theorem states that if any internal stress distribution can be found which does not exceed the yield stress at any point, and which is in equilibrium with the applied sustained loading, then the structure does not collapse.

Index